Plasma Nanoengineering and Nanofabrication

Special Issue Editors

Krasimir Vasilev
Melanie Ramiasa

MDPI • Basel • Beijing • Wuhan • Barcelona • Belgrade

MDPI

Special Issue Editors
Krasimir Vasilev
University of South Australia
Australia

Melanie Ramiasa
University of South Australia
Australia

Editorial Office
MDPI AG
St. Alban-Anlage 66
Basel, Switzerland

This edition is a reprint of the Special Issue published online in the open access journal *Nanomaterials* (ISSN 2079-4991) from 2015–2016 (available at: http://www.mdpi.com/journal/nanomaterials/special_issues/plasma_nano).

For citation purposes, cite each article independently as indicated on the article page online and as indicated below:

Author 1; Author 2. Article title. *Journal Name* **Year**, *Article number*, page range.

First Edition 2017

ISBN 978-3-03842-558-8 (Pbk)
ISBN 978-3-03842-559-5 (PDF)

Table of Contents

About the Special Issue Editors

Krasimir Vasilev is currently a Full Professor at the University of South Australia. He is one of the pioneers of the field of nanoengineered plasma polymer films and has developed innovative technologies such as antibacterial coatings, drug delivery platforms, cell guidance surfaces and diagnostic devices. He was awarded research funding in excess of 15 million dollars. He has published more than 160 research papers, reviews and book chapters. He is the inventor of five patents which underpin technologies currently under translation to commercialization.

Professor Vasilev was awarded four prestigious research fellowships i.e. Marie Curie Fellowship, Humboldt Fellowship for Experienced Researchers, ARC Future Fellowship and NHMRC Career Development Fellowship. Other honours and awards include the John A. Brodie Medal for achievements in Chemical Engineering in 2016 and the IAAM medal for contributions to the field of Advanced Materials in the year 2017.

Melanie Ramiasa has chemical engineering background and, in 2013, was awarded the Ian Wark Medal for her PhD thesis on dynamic wetting phenomena on nanorough surfaces. She has since been exploring the effects of surface chemistry and nanoscale topography on the interaction between (bio)materials and their environment. At the University of South Australia, he research focuses on developing bioengineered platforms for cell guidance substrates and cell capture devices for diagnostic purposes using both wet chemistry and plasma based methods.

Dr Macgregor is the South Australian representative for the Royal Australian Chemical Institute Polymer group. She has an emerging track record in the field of materials nanoengineering comprising of 3 patents and over 30 peer reviewed publications. She has led several competitive grant projects and been awarded various prizes such as the 2016 John A Brodie Medal and the 2017 UniSA ITEE division Early Career Researcher Award.

Preface to "Plasma Nanoengineering and Nanofabrication"

Plasma is a phenomenon defined as the forth state of matter. 99 per cent of the visible universe is plasma. The lightnings and the auroras are examples of natural plasmas that have intrigued, puzzled and fascinated humans for millennia. These plasmas are wild and uncontrollable. However, plasmas can also be confined, controlled and employed for generation of exciting new materials that benefit economy and society. Plasma driven processes such as etching and deposition revolutionized the semiconductor industry. High speed computer processors engineered at the nanoscale via plasma process have become inseparable part of our everyday life. New applications of plasmas in medicine and nanoscience are currently attracting attention with the promise to revolutionize many fields.

Surface engineering has been an area where plasma processes have traditionally made and continue making major impact. The ability to retain material bulk properties but only tune its properties at the surface in a desired fashion brings enormous value to numerous products in fields from medicine to electronics.

With the recent advances in nanotechnology, plasma nanofabrication has become an exciting new niche because plasma-based approaches can deliver unique structures at the nanoscale that cannot be achieved by other techniques and/or in a more economical and environmentally friendly manner. Over the last decade, there have been exiting breakthroughs in the utilization of plasma processes in the fabrication of a rich diversity of nanomaterials and nanoengineered coatings. Examples include nanowires, nanotubes, nanoparticles, and nanotextured coatings. Some of these materials can simply not be derived by conventional means while many others have remarkable plasma-induced properties setting them apart from their traditional analogue. These materials are revolutionizing a spectrum of fields, ranging from electronics to biology and medicine.

This book contains 11 original research and review type chapters from leading investigators in the field of plasma nanofabrication from across the world. The inspiring review by Puliyalil and Cvelbar focuses on the use of plasma etching for the production of advanced polymeric materials for the fields of microelectronic, photonics, water repellence and medical devices. It provides a comprehensive synopsis of the mechanism governing the selective plasma etching of polymeric materials and will without a doubt inspire new research towards the next generation of plasma-derived nanomaterials. Several other quality chapters also focus on plasma etching and surface nanotexturing. Fiflis et al. produced nanotexture palladium surface using helium plasma. These advanced material substrates exhibit great integrity and better catalytic properties than their smooth counterparts. In the work of Mangla et al. a dense plasma focus device was used to produce, for the first time, III–V semiconductors made of continuous and porous nano gallium arsenide (GaAs) composites. The interesting optical properties of these nanotextured materials (strong photoluminescence and high transmission) show potential for applications as transmission type photocathodes or in visible optoelectronic devices.

This book also contains chapters describing novel plasma based approaches for the synthesis of nanomaterials. Li et al. report on the first fabrication and characterisation of titanium alloy composite nanoparticles prepared via hydrogen plasma-metal reaction. Ananth and Mok used a non-thermal dielectric barrier discharged (DBD) atmospheric plasma to produce silver and ruthenium oxide nanocomposite. The materials prepared with this environmentally friendly low energy plasma method display high crystallinity and good thermal resistivity. Tanaka et al. describe the formation of nanoparticles by induction thermal plasma for four different lithium composite systems. This systematic study reveals the influence of melting temperature and boiling point of the source materials on the nucleation, growth and shapes of the nanoparticles produced.

The ability to homogeneously coat nanoparticles to control their outer surface properties is an important aspect of nanofabrication. This topic is tackled in this special issue by Post et al who describe a way to achieve homogeneous coating on metal nanoparticles and agglomerates using post-DBD.

Two chapters in this book are specifically focused on the practical applications of plasma derived nanomaterials and nanoengineered surfaces. Atmospheric pressure argon plasma was used to produce both silver and gold nanoparticles on tin oxide solar cells. In this application focused study, Dao and Dhoi demonstrated that the plasma fabricated nanoparticles can tune the efficiency of the solar cells. In another chapter Lai et al. use argon plasma treated zinc oxide nanowire to improve the resistive switching of single nanowire-based memory cells. The findings of this study will help the development of new-age, small size and transparent, resistive random access memory storage media.

This book also contains two modelling studies of plasma facilitated processes for the fabrication of nanomaterials. First, Xiong et al. investigated, from a theoretical point of view, the effect of key parameters such as droplet size, injection angle and velocity on nanoparticle processing in suspension plasma spray. The 3 dimensional model presented is validated with published experimental data and can be used to predict the nanoparticle suspension behaviour in novel spray coating technologies. In another chapter, Shigeta and Watanabe took a computational approach to model the effect of saturation pressure on the growth process and size-composition distribution of metal-silicide nanoparticles produced via thermal plasma. This study provides valuable theoretical insights on the very complex processes involved in the growth mechanisms of nanopowders.

The Editorial team would like to thank all contributing authors and the publisher for their support to make this book a success.

Krasimir Vasilev and Melanie Ramiasa

Special Issue Editors

nanomaterials

MDPI

Editorial

Plasma Nanoengineering and Nanofabrication

Krasimir Vasilev [1,*] and Melanie Macgregor Ramiasa [2]

[1] School of Engineering, University of South Australia, Adelaide, SA 5095, Australia
[2] Future Industries Institute, University of South Australia, Adelaide, SA 5095, Australia;
   melanie.ramiasa@unisa.edu.au
* Correspondence: krasimir.vasilev@unisa.edu.au; Tel.: +61-8-8302-5697; Fax: +61-8-8302-5689

Academic Editor: Thomas Nann
Received: 17 June 2016; Accepted: 20 June 2016; Published: 23 June 2016

With the recent advances in nanotechnology, plasma nanofabrication has become an exciting new niche because plasma-based approaches can deliver unique structures at the nanoscale that cannot be achieved by other techniques and/or in a more economical and environmentally friendly manner. Over the last decade, there have been exiting breakthroughs in the utilization of plasma processes in the fabrication of a rich diversity of nanomaterials and nanoengineered coatings. Examples include nanowires, nanotubes, nanoparticles, and nanotextured coatings. Some of these materials can simply not be derived by conventional means while many others have remarkable plasma-induced properties setting them apart from their traditional analogue. These materials are revolutionizing a spectrum of fields, ranging from electronics to biology and medicine.

The goal of this Special Issue is to present some of the latest advances in the fields of plasma assisted nanoengineering and nanofabrication and their application to modern technologies. In addition, the Special Issue aims at highlighting current challenges and obstacles that lie on the path to fully understand the fundamental physical phenomena underlying the plasma facilitated fabrication of nanomaterials. Further, this Special Issue intends to provide guidance to researchers in the field and inform the community of exciting future directions.

This special issue contains a blend of 11 original research papers, communications and review type contributions from leading investigators in the field of plasma nanofabrication from across the world. The inspiring review by Puliyalil and Cvelbar [1] focuses on the use of plasma etching for the production of advanced polymeric materials for the fields of microelectronic, photonics, water repellence and medical devices. It provides a comprehensive synopsis of the mechanism governing the selective plasma etching of polymeric materials and will without a doubt inspire new research towards the next generation of plasma-derived nanomaterials. Several other quality research articles published in this special issue also focus on plasma etching and surface nanotexturing. For Instance, Fiflis et al. [2] produced nanotexture palladium surface using helium plasma. These advanced material substrates exhibit great integrity and better catalytic properties than their smooth counterparts. In the work of Mangla et al. [3], a dense plasma focus device was used to produce, for the first time, III–V semiconductors made of continuous and porous nano gallium arsenide (GaAs) composites. The interesting optical properties of these nanotextured materials (strong photoluminescence and high transmission) show potential for applications as transmission type photocathodes or in visible optoelectronic devices.

This special issue also contains articles presenting a novel plasma based approach for the synthesis of nanomaterials. Li et al. [4] report on the first fabrication and characterisation of titanium alloy composite nanoparticles prepared via hydrogen plasma-metal reaction. Ananth and Mok [5] used gentle, non-thermal dielectric barrier discharged (DBD) atmospheric plasma to produce silver and ruthenium oxide nanocomposite. The materials prepared with this environmentally friendly low energy plasma method display high crystallinity and good thermal resistivity. Tanaka et al. [6]

investigated the formation of nanoparticles by induction thermal plasma for four different lithium composite systems. This systematic study reveals the influence of melting temperature and boiling point of the source materials on the nucleation, growth and shapes of the nanoparticles produced.

Another important aspect of plasma based nanomaterials, is to be able to homogeneously coat nanoparticles to control their outer surface properties. This topic is tackled in this special issue by Post et al. [7] who describe a way to achieve homogeneous coating on metal nanoparticles and agglomerates using post-DBD.

Two articles in this special issue are particularly focused on the practical applications of plasma derived nanomaterials and nanoengineered surfaces. Atmospheric pressure argon plasma was used to produce both silver and gold nanoparticles on tin oxide solar cells. In this application focused study, Dao and Dhoi [8] demonstrated that the plasma fabricated nanoparticles can tune the efficiency of the solar cells. Lai et al. [9] use argon plasma treated zinc oxide nanowire to improve the resistive switching of single nanowire-based memory cells. The findings of this study will help the development of new-age, small size and transparent, resistive random access memory storage media.

This special issue also presents two modelling studies of plasma facilitated processes for the fabrication of nanomaterials. First, Xiong et al. [10] investigated, from a theoretical point of view, the effect of key parameters such as droplet size, injection angle and velocity on nanoparticle processing in suspension plasma spray. The 3 dimensional model presented is validated with published experimental data and can be used to predict the nanoparticle suspension behaviour in novel spray coating technologies. Shigeta and Watanabe [11] took a computational approach to model the effect of saturation pressure on the growth process and size-composition distribution of metal-silicide nanoparticles produced via thermal plasma. This study provides valuable theoretical insights on the very complex processes involved in the growth mechanisms of nanopowders.

Finally, the Editorial team would like to thank all contributing authors for making this Special Issue a success.

Acknowledgments: The authors wish to acknowledge the support the University of South Australia Research Theme Investment Scheme 2015, Channel 7 Children Research Foundation (Project 15976) and the South Australian state Government Premier Research and Industry Fund.

References

1. Puliyalil, H.; Cvelbar, U. Selective Plasma Etching of Polymeric Substrates for Advanced Applications. *Nanomaterials* **2016**, *6*. [CrossRef]
2. Fiflis, P.; Christenson, M.P.; Connolly, N.; Ruzic, D.N. Nanostructuring of Palladium with Low-Temperature Helium Plasma. *Nanomaterials* **2015**, *5*. [CrossRef]
3. Mangla, O.; Roy, S.; Ostrikov, K. Dense Plasma Focus-Based Nanofabrication of III–V Semiconductors: Unique Features and Recent Advances. *Nanomaterials* **2016**, *6*. [CrossRef]
4. Li, J.Y.; Mei, Q.S. TiAl$_3$-TiN Composite Nanoparticles Produced by Hydrogen Plasma-Metal Reaction: Synthesis, Passivation, and Characterizatio. *Nanomaterials* **2016**, *6*. [CrossRef]
5. Ananth, A.; Mok, Y.S. Dielectric Barrier Discharge (DBD) Plasma Assisted Synthesis of Ag$_2$O Nanomaterials and Ag$_2$O/RuO$_2$ Nanocomposites. *Nanomaterials* **2016**, *6*. [CrossRef]
6. Tanaka, M.; Kageyama, T.; Sone, H.; Yoshida, S.; Okamoto, D.; Watanabe, T. Synthesis of Lithium Metal Oxide Nanoparticles by Induction Thermal Plasmas. *Nanomaterials* **2016**, *6*. [CrossRef]
7. Post, P.; Jidenko, N.; Weber, A.P.; Borra, J.-P. Post-Plasma SiO$_x$ Coatings of Metal and Metal Oxide Nanoparticles for Enhanced Thermal Stability and Tunable Photoactivity Applications. *Nanomaterials* **2016**, *6*. [CrossRef]
8. Dao, V.-D.; Choi, H.-S. Highly-Efficient Plasmon-Enhanced Dye-Sensitized Solar Cells Created by Means of Dry Plasma Reduction. *Nanomaterials* **2016**, *6*. [CrossRef]
9. Lai, Y.; Qiu, W.; Zeng, Z.; Cheng, S.; Yu, J.; Zheng, Q. Resistive Switching of Plasma-Treated Zinc Oxide Nanowires for Resistive Random Access Memory. *Nanomaterials* **2016**, *6*. [CrossRef]

10. Xiong, H.; Zhang, C.; Zhang, K.; Shao, X. Effects of Atomization Injection on Nanoparticle Processing in Suspension Plasma Spray. *Nanomaterials* **2016**, *6*. [CrossRef]
11. Shigeta, M.; Watanabe, T. Effect of Saturation Pressure Difference on Metal-Silicide Nanopowder Formation in Thermal Plasma Fabrication. *Nanomaterials* **2016**, *6*. [CrossRef]

3

nanomaterials

MDPI

Article

TiAl$_3$-TiN Composite Nanoparticles Produced by Hydrogen Plasma-Metal Reaction: Synthesis, Passivation, and Characterization

Ju Ying Li [1] and Qing Song Mei [2],*

[1]   School of Mechanical Engineering, Wuhan Polytechnic University, Wuhan 430023, China; jylimei@163.com
[2]   Department of Materials Engineering, School of Power and Mechanical Engineering, Wuhan University, Wuhan 430072, China
*   Correspondence: qsmei@whu.edu.cn; Tel.: +86-27-687-722-52

Academic Editor: Krasimir Vasilev
Received: 5 April 2016; Accepted: 23 May 2016; Published: 1 June 2016

Abstract: TiAl$_3$ and TiN composite nanoparticles were continuously synthesized from Ti–48Al master alloy by hydrogen plasma-metal reaction in a N$_2$, H$_2$ and Ar atmosphere. The phase, morphology, and size of the nanoparticles were studied by X-ray diffraction (XRD) and transmission electronic microscopy (TEM). X-ray photoelectron spectroscopy (XPS) and evolved gas analysis (EGA) were used to analyze the surface phase constitution and oxygen content of the nanoparticles. The as-synthesized nanopowders were mainly composed of nearly spherical TiAl$_3$ and tetragonal TiN phases, with a mean diameter of ~42 nm and mass fractions of 49.1% and 24.3%, respectively. Passivation in the atmosphere of Ar and O$_2$ for 24 h at room temperature led to the formation of amorphous Al$_2$O$_3$ shells on the TiAl$_3$ particle surface, with a mean thickness of ~5.0 nm and a mass fraction of ~23.5%, as well as TiO$_2$ with a mass fraction of ~3.2%.

Keywords: nanoparticles; TiAl$_3$; TiN; hydrogen plasma-metal reaction; passivation

1. Introduction

Hydrogen plasma-metal reaction (HPMR) is an effective method to synthesis nanoparticles of pure metals or alloys that was first developed by Uda and coauthors [1–3]. After that, improvements have been made on this method so that metal nanoparticles can be continuously produced [4]. The merits of HPMR include: (1) high generation rate; (2) wide applicability; (3) high purity of the produced particles; and (4) nanoscale particle size. By now, nanoparticles of pure metals and different binary or ternary alloys have been successfully produced by HPMR [5–9]. Synthesis of nanoparticles in a large quantity is important for the manufacturing of components by the traditional powder metallurgy method and the 3D rapid prototyping that has quickly developed in recent years [10]. The Ti-Al system is important for applications in automobile and aerospace industries. In previous studies [11, 12], nanoparticles of titanium aluminides were investigated in the Ti-Al binary alloy by HPMR. However, the synthesis of nanopowders containing Ti-Al intermetallic and ceramic nanoparticles has not been reported. Due to the large surface area, the as-synthesized metallic nanoparticles usually need to be passivated before full exposure to the air. Passivation of metallic nanoparticles (e.g., Al nanoparticles) was usually conducted in Ar and O$_2$ atmosphere, and sometimes in different liquid or solid substances [13,14]. As passivation can lead to changes in the composition, phase, and property of the surface of nanoparticles, the investigation and characterization of the surface of passivated nanoparticles are necessary [15–17] but quantitative characterization is usually difficult. In our previous study [18], Al$_2$O$_3$/Ti$_2$AlN composites with a novel combination of high temperature properties were fabricated successfully from TiAl$_3$-TiN composite nanoparticles by HPMR, for which quantitative characterization of the composition, surface structure, and phase fraction of the composite nanoparticles after passivation is of great importance.

In this study, we reported the synthesis and quantitative characterization of $TiAl_3$-TiN composite nanoparticles from a Ti-Al binary system by hydrogen plasma-metal reaction in a N_2, H_2, and Ar atmosphere. The phase, morphology, and size of the composite nanoparticles, as well as their passivation behaviors, were studied by X-ray diffraction (XRD), transmission electronic microscopy (TEM), X-ray photoelectron spectroscopy (XPS), and evolved gas analysis (EGA). The surface composition, structure, and phase fraction of the composite nanoparticles after passivation were determined.

2. Experiment Procedure

2.1. Synthesis of Nanoparticles

The master alloys used in this work were prepared from 99.5% sponge Ti and 99.8% Al buttons by melting three times using a consumable electrode vacuum furnace. The master alloy was designed as Ti–48Al (at. %, the same for below) and machined into 20 mm in diameter and 200 mm in height. The HPMR equipment used in this study was developed on the base of [1–3]. The chamber was then evacuated to about 100 Pa using a rotary pump, washed three times with argon gas, and backfilled with high purity argon, hydrogen, and nitrogen to a predetermined pressure. The chamber atmosphere is the mixture gas of N_2, H_2, and Ar (0.2:1:1) with a total pressure of 0.1 MPa. The master alloy underwent evaporation, reaction, and condensation to form nanoparticles. The nanoparticles were then transferred by a circulating pump and deposited onto the inner surfaces of the collection chamber. Passivation of nanoparticles was performed in the atmosphere of Ar and O_2 at room temperature for 24 h.

2.2. Characterization

XPS analysis was performed on the ESCALAD-250 spectrometer (Thermo Electron, Waltham, MA, USA) using monochromated Al Kα X-rays (1486.6 eV) and a hemispherical analyzer. The samples were mounted onto carbon adhesive tape. The operating parameters were as follows: the system base pressure was 1×10^{-6}–7×10^{-6} Pa; the diameter of the X-ray beam was 100 μm, and the angle of emission of the detected photoelectrons (relative to the surface normal) was 45°. The evolved gas analysis (EGA) was performed on a TC-436 Oxygen/Nitrogen determinator (LECO, Saint Joseph, MI, USA) operating in the inert gas fusion principle. The samples (0.1–0.3 g in mass) were mixed with a pre-degassed graphite powder as a reducing agent, placed into pre-degassed graphite crucibles, and ramp heated under a helium flow. The evolution of the CO and CO_2 gases was monitored on-line with non-dispersive infrared detectors (NDIR) (LECO, Saint Joseph, MI, USA). XRD was performed with an Ultima IV diffractometer (Rigaku, Tokyo, Japan) using Cu K$_\alpha$ radiation. The microscopic images of the nanoparticles were obtained using a JEOL-2000FX (JEOL, Tokyo, Japan) TEM. High resolution transmission electron microscopy (HRTEM) was conducted on a JEOL-TEM2100 transmission electron microscope (JEOL, Tokyo, Japan). The sample was sonicated in acetone and dropped onto a carbon coated copper grid.

3. Results and Discussion

Figure 1 shows typical bright-field TEM images of nanoparticles produced from Ti–48Al master alloy in the N_2, H_2, and Ar atmosphere. Two kinds of morphologies of the as-synthesized nanoparticles can be seen from Figure 1: nearly spherical and tetragonal shapes, all dispersing well on the carbon film. The inset of Figure 1b is the corresponding micro-diffraction pattern of the tetragonal particle, which can be indexed as TiN phase. At high temperatures produced by the electric arc, nitrogen atoms can react with the metal vapor containing two elements: Ti and Al. Here TiN phase is formed instead of AlN by the selective reaction of N and Ti atoms. Thermodynamic analysis can explain this selectivity: at the same temperature, the enthalpy of formation of TiN is lower than that of AlN (e.g., the values of enthalpy of formation of TiN and AlN at 298 K are −339.4 kJ/mol and −319.2 kJ/mol, respectively, and those at 2300 K are −264.9 kJ/mol and −258.6 kJ/mol, respectively) [19]. Figure 1c is the particle size distribution of as-synthesized nanoparticles. As shown in Figure 1c, the distribution of nanoparticle size is between 10 nm and 200 nm, with an average of about 42 nm. Interestingly, although the as-synthesized nanoparticles contain two main phases with different morphologies, the average sizes of them are similar. The XRD pattern (Figure 2) further indicates that the as-synthesized nanoparticles comprise two main phases of $TiAl_3$ and TiN. This is different from the Ti-Al nanoparticles synthesized in Ar and H_2 atmosphere: the Ti-Al nanoparticles synthesized in Ar and H_2 atmosphere from the same master alloy are composed of $TiAl_3$, Ti_2Al_5, TiAl, and Al phases [11]. In this study, TiAl and Ti_2Al_5 phases did not appear as indicated by XRD. This is because in the nitrogen-containing atmosphere, Ti first reacts with N to form TiN, which consumes a large fraction of Ti. As Ti content in the vapor decreases, the formation of TiAl and Ti_2Al_5 phases is suppressed.

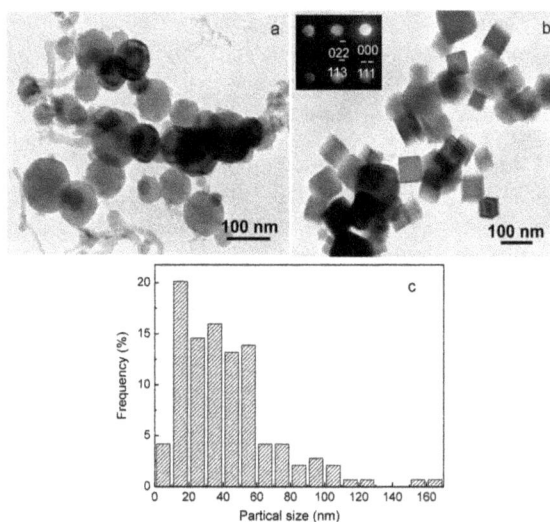

Figure 1. (**a**) and (**b**) Typical bright-field transmission electron microscope (TEM) micrographs of the nanoparticles synthesized from the master alloy of Ti–48Al by hydrogen plasma-metal reaction (HPMR) in N_2, H_2, and Ar atmosphere; (**c**) particle size distribution.

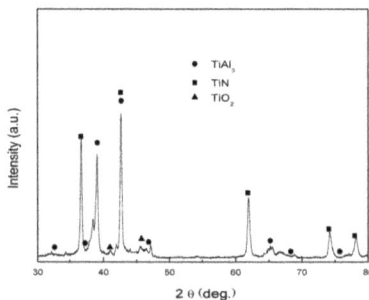

Figure 2. X-ray diffraction analysis (XRD) pattern of the nanoparticles synthesized from the master alloy of Ti–48Al by HPMR in N_2, H_2, and Ar atmosphere.

Due to the large surface area and high surface activity, the composite nanoparticles were passivated to avoid violent reaction with oxygen by direct exposure to air. Figure 3 is the XPS result of the composite nanoparticles after passivation. As shown in Figure 3, the surface elements of nanopowders are mainly composed of Al, O, and Ti. As can be seen from Figure 3b,c, the O_{1S} photoelectron spectrum shows the binding energy of 531.5 eV, and the Ti_{2p} photoelectron spectrum shows the binding energy of 458.8 eV. As shown in Figure 3d, the fitting analysis of the Al_{2p} photoelectron spectrum agrees well with the experiment data, indicating the binding energies of 73.2 eV and 75.6 eV, respectively. Compared with the standard database, the state of Al element is 3+ of Al_2O_3, and the state of Ti element is 4+ of TiO_2 [20]. A semi quantitative analysis of the surface elemental composition of the composite nanoparticles is listed in Table 1. Table 1 generally indicates that the main surface phase of nanopowders is Al_2O_3, with only a small amount of TiO_2.

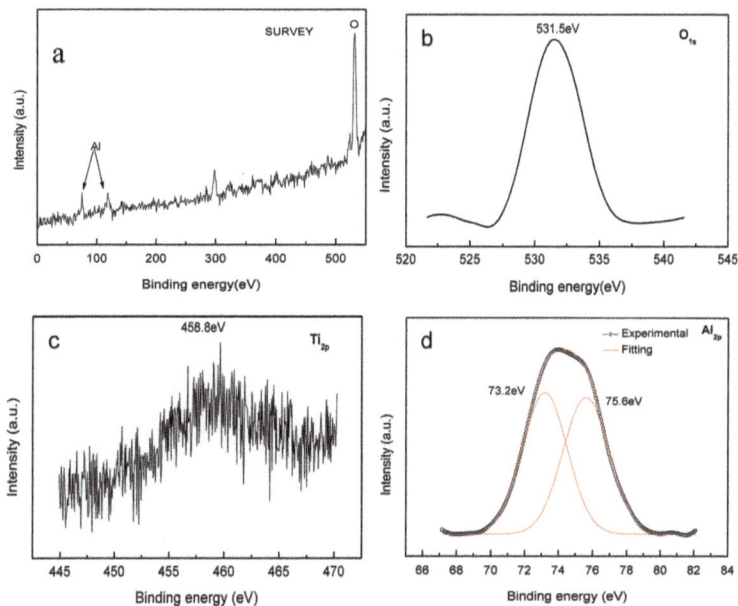

Figure 3. (**a**) X-ray photoelectron spectroscopy (XPS) patterns of the nanoparticles synthesized from the master alloy of Ti–48Al by HPMR in N_2, H_2, and Ar atmosphere; (**b–d**) are the enlargements corresponding to different ranges of binding energy.

Table 1. Surface composition of the nanopowders.

Element	O	Al	Ti
Content (at. %)	60.9%	38.4%	1.7%

Figure 4 shows the total oxygen release curve of the passivated nanoparticles. As shown in Figure 4a, the main oxygen release peak occurs at the time between 110 s and 152 s, and the temperature between 1900 °C and 2100 °C, which corresponds to the oxygen release peak of Al_2O_3 phase in the nanopowders. The corresponding oxygen fraction for this peak is 12.22 wt. %. As shown in Figure 4b, the oxygen release peak at the time between 66 s and 110 s and the temperature between 1400 °C and 1800 °C has the oxygen content of 1.30 wt. %. It can be concluded that this is the oxygen release peak for TiO_2 in the nanopowders. In Figure 4b, the oxygen release peak at the time between 0 s and 41 s and the temperature between 2200 °C and 1100 °C has the oxygen content of 0.18 wt. %, which corresponds to the release peak of adsorbed oxygen. The above results indicate that oxygen in the nanopowders mainly exists in the form of Al_2O_3 compounds, with a small amount in TiO_2 compounds and surface adsorbed oxygen.

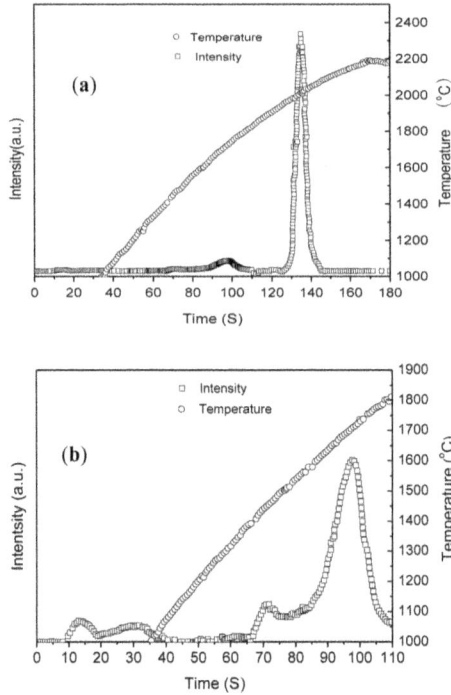

Figure 4. (a) Release curve of oxygen in the passivated nanopowders synthesized from the master alloy of Ti–48Al by HPMR in N_2, H_2, and Ar atmosphere; (b) the amplification in the time range of 0–110 s.

The mass fraction of TiN (W_{TiN}) in the powder can be calculated by:

$$W_{TiN} = W_N + W_{Ti/TiN} \tag{1}$$

where W_N is the mass fraction of N in the powder, and $W_{Ti/TiN}$ is the mass fraction of Ti in TiN, which can be calculated by:

$$W_{Ti/TiN} = W_N(M_{Ti} + M_N)/M_N \tag{2}$$

where M_{Ti} and M_N are the atomic mass of Ti and N, respectively. The mass fraction of TiO_2 (W_{TiO_2}) in the powder can be calculated by:

$$W_{TiO_2} = W_{O/TiO_2} + W_{Ti/O_2} \tag{3}$$

where W_{O/TiO_2} and W_{Ti/TiO_2} are the mass fraction of O and Ti in TiO_2, respectively. W_{Ti/TiO_2} can be calculated by:

$$W_{Ti/TiO_2} = W_{O/TiO_2}M_{Ti}/2M_O \tag{4}$$

where M_O is the atomic mass of oxygen.

Similarly, the mass fraction of Al_2O_3 ($W_{Al_2O_3}$) in the powder can be calculated by:

$$W_{Al_2O_3} = W_{O/Al_2O_3}(1 + 2M_{Al}/3M_O) \tag{5}$$

where M_{Al} is the atomic mass of Al. Also, the mass fraction of $TiAl_3$ (W_{TiAl_3}) can be calculated by:

$$W_{TiAl_3} = (W_{Ti} - W_{Ti/TiN} - W_{Ti/TiO_2}) \times (1 + 3M_{Al}/M_{Ti}) \qquad (6)$$

where W_{Ti} is the mass fraction of Ti in the powder.

Table 2 lists the elemental composition of the nanopowders after passivation, measured by gas analysis and chemical analysis, as well as the fractions of oxygen in Al_2O_3, TiO_2, and adsorbed oxygen. Using the data in Table 2, the mass fraction of difference phases in the nanopowders can be calculated by Equations (1)–(6), and the results are given in Table 3.

Table 2. Mass fractions of the elements in the nanopowders and those of oxygen in different phases. See the text for details.

W_N	W_{Ti}	W_{Al}	W_O		
			W_{OA}	W_{O/Al_2O_3}	W_{O/TiO_2}
5.5%	39.0%	41.8%	0.18%	12.22%	1.3%

Table 3. Calculated mass fractions of different phases in the nanopowders. See the text for details.

W_{TiN}	W_{TiAl_3}	$W_{Al_2O_3}$	W_{TiO_2}
24.3%	49.1%	23.5%	3.2%

As shown in Table 3, Al_2O_3 has a mass fraction of about 23.5% in the nanopowders. However, the XRD pattern of the nanopowders (Figure 2) does not show the Al_2O_3 peak. This is because the surface oxide is an amorphous structure, as shown in Figure 5. As shown in Figure 5, Al_2O_3 forms a thin shell structure on the surface of the $TiAl_3$ particle, as similar to the surface oxide shells of Al and Fe nanoparticles [16,17]. Here, the average shell thickness (t) can be estimated by:

$$t = \left(\sqrt[3]{1 + \frac{W_{Al2O3}\,\rho_{TiAl3}}{W_{TiAl3}\,\rho_{Al2O3}}} - 1 \right) R \qquad (7)$$

where R is the radius of the $TiAl_3$ particle, and $\rho_{Al_2O_3}$ and ρ_{TiAl_3} are the density of $TiAl_3$ and Al_2O_3, respectively. From Equation (7), an average shell thickness of about 5.0 nm is obtained, which agrees well with the HRTEM observations (Figure 5). The surface oxide Al_2O_3 is dense and protective, which can hinder the further oxidation of nanopowders in air. No further increase of the oxygen content was found in the sample of nanopowders after exposed in air for more than 600 h.

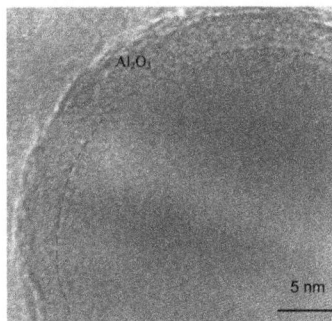

Figure 5. High resolution transmission electron microscopy (HRTEM) micrograph of the passivated $TiAl_3$ nanoparticle showing the surface oxide.

4. Conclusions

In this study, nanopowders were produced continuously by the HPMR method in a N_2, H_2 and Ar atmosphere from the master alloy of Ti–48Al, followed by passivation in the Ar and O_2 atmosphere for 24 h at room temperature. The phase constitution, morphology, and size of the nanopowders were investigated, and the surface composition and phase were quantitatively characterized. The main results are summarized as follows:

(1) The nanopowders are mainly composed of $TiAl_3$ and TiN phases with an average diameter of ~42 nm. $TiAl_3$ nanoparticles are nearly spherical with a mass fraction of ~49.1%, and TiN nanoparticles are tetragonal with a mass fraction of ~24.3%.

(2) Passivation of the nanopowders led to the formation of protective amorphous Al_2O_3 shells on the particle surface, with a mean thickness of ~5.0 nm and a mass fraction of ~23.5%, as well as TiO_2 with a mass fraction of ~3.2%.

Acknowledgments: Financial supports from the National Natural Science Foundation of China (Grant 51371128) and the Research Foundation of Wuhan Polytechnic University (Grant 2014833) are acknowledged. Q.S.M. is also supported by the Research Foundation of Wuhan University.

Author Contributions: Q.S.M. and J.Y.L. conceived and designed the experiments; J.Y.L. performed the experiments; Q.S.M. and J.Y.L. analyzed the data and wrote the paper.

Conflicts of Interest: The authors declare no conflict of interest.

References

1. Uda, M. Synthesis of new ultrafine metal particles. *Bull. Jpn. Inst. Met.* **1983**, *22*, 412–420. [CrossRef]
2. Ohno, S.; Uda, M. Generation rate of ultrafine metal particles in hydrogen plasma-metal reaction. *J. Jpn. Inst. Met.* **1984**, *48*, 640–646.
3. Ohno, S.; Uda, M. Preparation for ultrafine particles of Fe–Ni, Fe–Cu and Fe–Si alloys by hydrogen plasma-metal reaction. *J. Jpn. Inst. Met.* **1999**, *53*, 946–952.
4. Cui, Z.L.; Zhang, Z.K.; Hao, C.C.; Dong, L.F.; Meng, Z.G.; Yu, L.Y. Structures and properties of nano-particles prepared by hydrogen arc plasma method. *Thin Solid Films* **1998**, *318*, 76–82. [CrossRef]
5. Li, X.G.; Liu, T.; Sato, M.; Takahashi, S. Synthesis and characterization of Fe–Ti nanoparticles by nitrogen plasma metal reaction. *Powder Technol.* **2006**, *163*, 183–187. [CrossRef]
6. Ohsaki, K.; Uda, M.; Okazaki, K. Preparation of ultra-fine titanium powder by arc-plasma process. *Mater. Trans. JIM* **1995**, *36*, 1386–1391. [CrossRef]
7. Liu, T.; Zhang, Y.H.; Li, X.G. Preparations and characteristics of Ti hydride and Mg ultrafine particles by hydrogen plasma–metal reaction. *Scr. Mater.* **2003**, *48*, 397–402. [CrossRef]
8. Li, X.G.; Takahashi, S. Synthesis and magnetic properties of Fe–Co–Ni nanoparticles by hydrogen plasma-metal reaction. *J. Magn. Magn. Mater.* **2000**, *214*, 195–203. [CrossRef]
9. Sakka, Y.; Okuyama, H.; Uchikoshi, T.; Ohno, S. Synthesis and characterization of Fe and composite Fe–TiN nanoparticles by dc arc-plasma. *J. Alloys Compd.* **2002**, *346*, 285–291. [CrossRef]
10. Murr, L.E.; Gaytan, S.M.; Ceylan, A.; Martinez, E.; Martinez, J.L.; Hernandez, D.H.; Machado, B.I.; Ramirez, D.A.; Medina, F.; Collins, S.; *et al.* Characterization of titanium aluminide alloy components fabricated by additive manufacturing using electron beam melting. *Acta Mater.* **2010**, *58*, 1887–1894. [CrossRef]
11. Li, J.Y.; Mei, Q.S. Synthesis of nanoparticles of titanium aluminides by hydrogen plasma-metal reaction: Effects of master alloy composition and chamber pressure on particle size, composition and phase. *Powder Metall.* **2015**, *58*, 209–213. [CrossRef]
12. Luo, J.S.; Li, K.; Li, X.B.; Shu, Y.J.; Tang, Y.J. Phase evolution and alloying mechanism of titanium aluminide nanoparticles. *J. Alloys Compd.* **2014**, *615*, 333–337. [CrossRef]
13. Kwon, Y.S.; Gromov, A.A.; Strokova, J.I. Passivation of the surface of aluminum nanopowders by protective coatings of the different chemical origin. *Appl. Surf. Sci.* **2007**, *253*, 5558–5564. [CrossRef]
14. Nazarenko, O.B.; Amelkovich, Y.A.; Sechin, A.I. Characterization of aluminum nanopowders after long-term storage. *Appl. Surf. Sci.* **2014**, *321*, 475–480. [CrossRef]

15. Krasovskii, P.V.; Samokhin, A.V.; Malinovskaya, O.S. Characterization of surface oxide films and oxygen distribution in α-W nanopowders produced in a DC plasma reactor from an oxide feedstock. *Powder Technol.* **2015**, *286*, 144–150. [CrossRef]

16. Fung, K.K.; Qin, B.X.; Zhan, X.X. Passivation of α-Fe nanoparticle by γ-Fe_2O_3 epitaxial shell. *Mater. Sci. Eng. A* **2000**, *286*, 135–138. [CrossRef]

17. Sánchez-López, J.C.; Caballero, A.; Fernkndez, A. Characterisation of Passivated Aluminium Nanopowders: An XPS and TEM/EELS Study. *J. Eur. Ceram. Soc.* **1998**, *18*, 1195–1200. [CrossRef]

18. Li, J.Y.; Mei, Q.S.; Cui, Y.Y.; Yang, R. Production of Al_2O_3/Ti_2AlN composite with novel combination of high temperature properties. *Mater. Sci. Eng. A* **2014**, *607*, 6–9. [CrossRef]

19. Haynes, W.M. *Handbook of Chemistry and Physics*, 96th ed.; Bruno, T.J., Lide, D.R., Eds.; CRC Press: Boca Raton, FL, USA, 2016.

20. Moulder, J.F. *Handbook of X-ray Photoelectron Spectroscopy*; Wagner, C.D., Riggs, W.M., Davies, L.E., Muilenberg, G.E., Eds.; Perkin-Elmer: Waltham, MA, USA, 1979.

nanomaterials

MDPI

Article

Effects of Atomization Injection on Nanoparticle Processing in Suspension Plasma Spray

Hong-bing Xiong [1,*], Cheng-yu Zhang [1], Kai Zhang [1,2] and Xue-ming Shao [1,*]

[1] Key Laboratory of Soft Machines and Smart Devices of Zhejiang Province, Zhejiang University, Hangzhou 310027, China; zcy330106@outlook.com (C.Z.); zhangk_0079@163.com (K.Z.)

[2] Research Institute of Petroleum Engineering, Shengli Oilfield, Sinopec, Dongying 257000, China

[*] Correspondence: hbxiong@zju.edu.cn (H.X.); mecsxm@zju.edu.cn (X.S.); Tel.: +86-571-87952200 (H.X.); +86-571-87951768 (X.S.)

Academic Editor: Thomas Nann
Received: 4 February 2016; Accepted: 11 May 2016; Published: 20 May 2016

Abstract: Liquid atomization is applied in nanostructure dense coating technology to inject suspended nano-size powder materials into a suspension plasma spray (SPS) torch. This paper presents the effects of the atomization parameters on the nanoparticle processing. A numerical model was developed to simulate the dynamic behaviors of the suspension droplets, the solid nanoparticles or agglomerates, as well as the interactions between them and the plasma gas. The plasma gas was calculated as compressible, multi-component, turbulent jet flow in Eulerian scheme. The droplets and the solid particles were calculated as discrete Lagrangian entities, being tracked through the spray process. The motion and thermal histories of the particles were given in this paper and their release and melting status were observed. The key parameters of atomization, including droplet size, injection angle and velocity were also analyzed. The study revealed that the nanoparticle processing in SPS preferred small droplets with better atomization and less aggregation from suspension preparation. The injection angle and velocity influenced the nanoparticle release percentage. Small angle and low initial velocity might have more nanoparticles released. Besides, the melting percentage of nanoparticles and agglomerates were studied, and the critical droplet diameter to ensure solid melting was drawn. Results showed that most released nanoparticles were well melted, but the agglomerates might be totally melted, partially melted, or even not melted at all, mainly depending on the agglomerate size. For better coating quality, the suspension droplet size should be limited to a critical droplet diameter, which was inversely proportional to the cubic root of weight content, for given critical agglomerate diameter of being totally melted.

Keywords: suspension plasma spray; atomization; nanoparticles; multiphase flow; thermal spray

1. Introduction

The technology of suspension plasma spray (SPS) is a novel spray technology [1]. In the SPS process, a liquid feedstock is used to inject nanometer-sized particles with the aid of a suspension. The suspension mixes the solid nanoparticles and the solution of water or alcohol, where the nanoparticles are usually agglomerated due to their high surface activity. Liquid feedstock spraying in general could offer unique opportunities in designing and fabricating complex material architectures with controlled and hierarchical microstructures. For example, the thermoelectric modules and solar cells were recently made from thermal sprayed silicon wafers [2]. Further, liquid feedstock spraying could lead to advancement of the spraying industry to spray nanoparticles in order to obtain dense and thick coating with good bond strength.

Suspension plasma spray process involves a series of complex phenomena, such as feedstock injection, suspension breakup and evaporation, and nanoparticles or their agglomerates release into the plasma jet [3]. The solid particles will experience further heating, melting, and evaporation before impacting onto the substrate. Researchers are attempting to analyze the above-mentioned complex phenomena in order to suggest the link between them. Modeling and numerical methods are employed to better understand the flow physics related to suspension atomization and spray evolution.

Though many experimental investigations have been conducted concerning SPS spray process [1], fundamental understanding lags behind the applications, and there is limited research concerning droplet and particle behaviors. Marchand *et al.* [4] studied the influence of liquid injection angle, and results showed that the relative velocity between liquid drop size, drop flow and flow field and the surface tension of liquid drop have significant influence on the incident and fracture of the liquid. Huang *et al.* [5] adopted a stochastic approach to investigate the particle behavior with two-fluid parcels, and found that the existence of large-scale eddies, variable property, Knudsen effect and mass-transfer cooling effect would affect the particle motion and heating history. Fazilleau [6] found that the fragmentation and vaporization of suspension or droplets occur about 10 to 15mm downstream of the injection location, and solvent vaporization cools down the plasma jet. Ozturk and Cetegen [7] presented a physical model to analyze the droplet evaporation and particle behavior during plasma spraying. Jabbari *et al.* [8] used FLUENT software to simulate the liquid fragmentation without considering the droplet evaporation. Although some particular aspects of droplet and particle behavior have been studied, there is still a lack of comprehensive model that connects the suspension droplets to the nanoparticles. In addition, agglomerate sintering always occurs in SPS due to insufficient atomization and might degrade the nano-structured coating quality. Such effect is important but not well considered in these studies.

In this paper, a comprehensive three-dimensional model is presented to examine the suspension droplets and spray evolution after liquid atomization. Two sub-models are used. First, the primary breakup of annular liquid sheath is calculated by a one-dimensional model based on Lund [9]. Second, sub-model is the Eulerian/Lagrangian modeling of the plasma jet and injected droplets with suspended nanoparticles. The first sub-model provides the proper initial and boundary conditions for the second sub-model. Then, each atomized droplet motion and heating can be tracked in the three-dimensional geometry, considering the nanoparticles release and melting. After validation with published experimental data, this model is used to predict the droplet and particle evolution under different operating conditions. The influence of operating conditions on the nanoparticle release and melting are discussed.

2. Mathematic Model

As experiments observed, the flow structure at the near nozzle exit and the downstream jet, primary breakup and secondary breakup occur in different stages to fragmentize the liquid to droplets. In the following Section 2.1, three steps will be shown on how to calculate the primary breakup. After that, Section 2.2 will describe how the droplets experience further breakup in the high-velocity plasma jet, as well as other physical particle phenomena of acceleration, heating, melting and evaporation.

2.1. Modeling Liquid Primary Breakup to Estimate Droplet Mean Size

Liquid primary breakup arises at the nozzle exit. There are three steps to fragmentize the continuous liquid into individual droplets. Firstly, the annular liquid sheath breaks up into a number of cylindrical ligaments, as indicated in Figure 1. Secondly, the ligaments breakup to ligament fragments. Finally, the ligament fragments stabilize to drops. To model droplet primary breakup, a one-dimensional primary breakup model, based on the one of Lund *et al.* [9], is described in this study and used to estimate the droplet mean size from the atomizing gas and liquid mass flow rates, liquid physical properties, and atomizer exit geometry. It is assumed that the annular liquid sheath breaks up to several cylindrical ligaments of almost same diameter as the thickness of the annular sheath. The ligaments further break up to ligament fragments at the wavelength of the most rapidly growing wave. Each fragment is assumed to form one droplet.

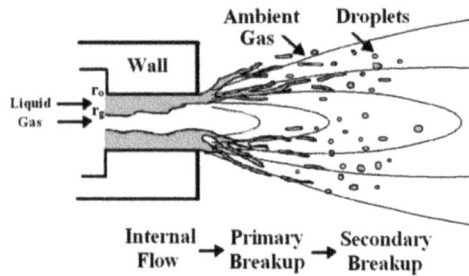

Figure 1. The schematic of liquid atomization.

Estimation of annular liquid sheath thickness: To estimate the thickness of annular liquid sheath, a simple model is adopted. It is assumed that the annular two-phase flow within the discharge orifice is one-dimensional, in viscous and isothermal, with compressible ideal gas and small interface velocity slip ratio. The velocity of the gas flow, v_g, satisfies the momentum equation as:

$$\frac{dp}{\rho_g} + v_g dv_g = 0 \tag{1}$$

Using the state of relation, the integration of above equation gives out:

$$RT\ln\left(\rho_g RT\right) + \frac{1}{2}\left(\frac{\dot{m}_l ALR}{\rho_g \pi r_g^2}\right)^2 = cosnt \tag{2}$$

where r_g is the radius of gas flow, ρ_g, the gas density, $\dot{m}_l$ is the mass flow rate of liquid and ALR is the air–liquid ratio by mass. The radius of gas flow can be written in terms of orifice radius, r_o, using the definition of void fraction α as: $r_g = \sqrt{\alpha}r_o$. According to Ishii [10], the interface velocity slip ratio "sr" under different flow rate can be expressed as:

$$sr = \sqrt{\frac{\rho_l}{\rho_g}\frac{\sqrt{\alpha}}{1 + C(1 - \alpha)}} \tag{3}$$

where C is the experimental coefficient. The interface velocity slip ratio and void fraction also has the relationship of:

$$1 + \frac{\rho_g sr}{\rho_l ALR} = \frac{1}{\alpha} \tag{4}$$

By solving Equations (2)–(4), ρ_g, α and sr can be calculated for different operating conditions. The thickness of annular liquid sheath is then calculated as: $\delta = r_o - r_g$, which is also the diameter of the typical cylindrical ligament.

Droplet mean size from primary breakup: After obtaining the thickness of annular liquid sheath, we need to compute the length of a typical ligament fragment, *i.e.*, the wavelength λ at which the disturbance grows most rapidly. Among the linear instability analysis theories, the one given by Weber is mostly widely used to determine λ [11]:

$$\lambda = \sqrt{2}\pi\delta\sqrt{1 + \frac{3\mu_l}{\sqrt{\rho_l \sigma_l \delta}}} \tag{5}$$

However, this equation does not consider movement of the sheet and is approximately correct only in the case of long wave disturbances. A more accurate solution provided by Senecal *et al.* [12] is utilized, in which the growth rate of surface disturbances for the sinusoidal mode is given by:

$$\omega = -\frac{2\mu_l k^2 \tanh(kh)}{\tanh(kh)+Q} + \frac{\sqrt{4\mu_l^2 k^4 \tanh^2(kh) - Q^2 V^2 k^2 - [\tanh(kh)+Q](-QV^2 k^2 + \sigma_l k^3/\rho_l)}}{\tanh(kh)+Q} \tag{6}$$

where k is the wave number given by $k = 2\pi/\lambda$, h is the half-thickness of sheath $h = \delta/2$, Q is the gas/liquid density ratio, V is the liquid sheath velocity, and μ_l, ρ_l, and σ_l represent the liquid viscosity, density, and liquid surface tension, respectively. Here, we use the relative velocity difference of the gas and the liquid at the nozzle exit. The breakup wavelength λ is determined by the wave number k where the maximum growth rate ω occurs in the curve calculated by Equation (6). Assuming that each fragment stabilizes to one droplet, the Sauter Mean diameter of drop size can then be calculated from the conservation of mass:

$$\text{SMD} = \left[\frac{3}{2}\delta^3\lambda\right]^{1/3} \tag{7}$$

Using this model, droplets mean diameter after primary breakup can be calculated and the initial droplet diameter is calculated based on this SMD diameter with a Rosin–Rammler distribution.

2.2. Eulerian/Lagrangian Model of Droplets or Particles in Plasma Jet

The droplets with imbedded solid particles experienced accelerating and heating in the plasma jet, and a code named LAVA-P-3D [13] was developed to simulate these processes as shown in Figure 2. The plasma gas flow fields were obtained by solving the Navier-Stokes (N-S) equations in Eulerian method, and tracking the particles as Lagrangian entities. The N-S equations of the plasma jet were established by assuming that the plasma jets were continuum, multi-component, compressible and chemically reactive, with temperature-dependent transport properties in the local thermodynamic equilibrium. The turbulence was simulated using k-ε model. Details of the plasma jet simulation could be found in reference [13].

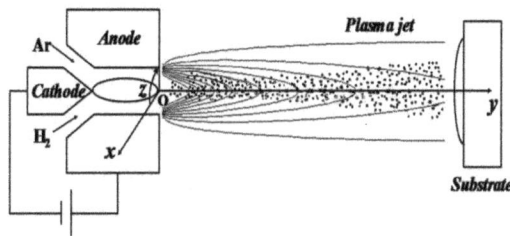

Figure 2. The schematic of suspension plasma spray with axially injected droplets or particles.

For the droplets injected into the plasma jet, their evolution include droplet secondary breakup, solvent evaporation, particle acceleration, heating and melting. These behaviors were simulated during their flight. In plasma flow field, the volume fraction of particles is less than 10^{-4}, so the collisions between particles were neglected [13].

Droplet secondary breakup modeling: When the droplets with initial diameter estimated from Equation (7) were injected into the plasma jet, secondary atomization would continually breakup the droplets, especially near the orifice with strong aerodynamic interaction. To model the droplet secondary breakup, a cascade atomization and droplet breakup (CAB) model as shown in Figure 3 [14] has been utilized, which is determined by the gas aerodynamic force, the liquid viscosity, and the surface tension force.

Figure 3. Schematic of droplet secondary breakup.

In this model, the droplet distortion is described by the deformation parameter, $y = 2x/r$, where x denotes the radial cross-section change from its equilibrium position and r is the drop radius. The deformation parameter is calculated by [15]:

$$\ddot{y} + \frac{5\mu_l}{\rho_l a^2}\dot{y} + \frac{8\sigma}{\rho_l a^3}y = \frac{2\rho_g |U|^2}{3\rho_l a^2} \tag{8}$$

where ρ denotes the density, μ the viscosity, σ the surface tension, U the relative drop-gas velocity, and the subscripts g and l denote the gas or liquid properties, respectively. Drop breakup occurs when the normalized drop distortion, $y(t)$, exceeds the critical value 1.

The creation of the product droplets is derived by using elements from population dynamics: for each breakup event it is assumed that the number of product droplets is proportional to the number of critical parent drops, where the proportionality constant depends on the drop breakup regime. From this, one can define the rate of droplet creation, which, in conjunction with the mass conservation principle between parent and product droplets, leads to the basic cascade breakup law:

$$\frac{d}{dt}\overline{m}(t) = -3K_{bu}\overline{m}(t) \tag{9}$$

where $m(t)$ denotes the mean mass of the product drop distribution, and the breakup frequency K_{bu} depends on the drop breakup regimes. As suggested by Reitz [16], three breakup regimes are classified with respect to increasing gas Weber number, bag breakup, stripping breakup regime, and catastrophic breakup regime. However, in this study, gas Weber number is mostly lower than 80, which falls into bag breakup regime, as shown in Figure 3. Here, Weber number provides the importance of inertia compared to surface tension. Breakup frequency $K_{bu} = 0.05\omega$ as suggested by O'Rourke and Amsden [15] is used in this study, and the drop oscillation frequency ω is given by

$$\omega^2 = \frac{8\sigma}{\rho_l a^3} - \frac{25\mu_l^2}{4\rho_l^2 a^4} \tag{10}$$

Note that, except for the mean mass $m(t)$, the actual size distribution of the product droplet has not been specified yet. For the model implementation in this study, a uniform product drop size distribution has been assumed, by which Equation (9) becomes:

$$\frac{r}{a} = e^{-K_{bu}t_{bu}} \tag{11}$$

where a and r are the radii of the parent and product drops, respectively, and t_{bu} is the breakup time, *i.e.*, the time until the normalized deformation $y(t)$ in the solution of Equation (8) exceeds the value of 1.

Solvent evaporation of droplets: The suspension droplets would evolve to agglomerates and nanoparticles due to solvent evaporation. As shown in Figure 4, the nano-sized solid particles were initially suspended in the micro-sized droplets, with small amount of particle aggregation, and uniform distribution. The aerodynamic interaction between the plasma gas and the droplet would further breakup the droplets into smaller pieces. At the same time, the hot plasma gas would also heat up and vaporize the solvent inside droplets. Once the solvent in droplets was totally vaporized out, the gas would blow away the solids to form individual nanoparticles. Otherwise, if the gas velocity is not large enough, the remaining solids would be sintered by the hot gas and the micro-size agglomerates would form thereafter. In this study, the nanoparticle size is larger than 1 nm but less than 1 μm, and the agglomerate is larger than or equal to 1 μm. These agglomerates or nanoparticles are simulated as new Lagrangian entities with their parent particle's position, velocity and temperature.

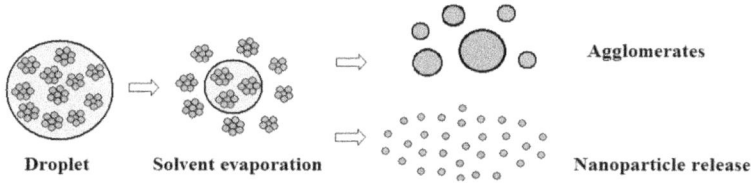

Figure 4. Schematic of droplet injection, solvent evaporation and nanoparticle release. Reproduced with permission from [17]. Copyright Courtesy of Delbos, 2006.

The droplet temperature can be expressed by:

$$T_d = T_{d,0} + \frac{Q_d}{m_d c_{p,d}} t \; if \; T_d < T_{m,d}$$
$$T_d = T_{m,d} \; if \; m_d c_{p,d} \left(T_{m,d} - T_{d,0}\right) \leqslant Q_d \leqslant m_d c_{p,d} \left(T_{m,d} - T_{d,0}\right) + m_d \alpha_{sl} L_{v,sl} \tag{12}$$

where $T_{d,0}$, Q_d, m_d and $c_{p,d}$ are the initial droplet temperature, heat gain, mass, and the specific heat of the droplet, respectively. Q_d can be calculated by $Q_d = Q_{conv} - Q_{rad}$, where Q_{conv} represents the convection heat, and Q_{rad} is the radiation heat loss of particles. $c_{p,d}$ is calculated based on the average of the mass fraction of solid particle and solvent as: $c_{p,d} = c_{p,p} \left(1 - \alpha_{sl}\right) + c_{p,sl} \alpha_{sl}$.

When the droplet solvent is totally vaporized, the solid particles contained in the droplet will be released into the plasma jet. They might be released as micrometer-sized agglomerates containing many nano-sized particles or individual nano-sized particles due to the evaporation of solvent and the aerodynamic forces of the plasma gas [18]. These individual micro- or nano-sized particles are treated as new Lagrangian entities with the current parameters of their mother particle, including position, droplet velocity, and temperature.

Particle acceleration and tracking model: For particles in plasma jets, the forces imparted on the particles are mainly the drag force, Saffman lift force and Brownian force. For particles smaller than 100 μm, the drag force is prominent. For the particles near the jet edge and the substrate, where the flow shear stress is large, the Saffman lift force is significant. While for the sub-micron or nanoparticles, Brownian force is important. By accounting for these three forces, the particles acceleration rate could be expressed as,

$$\overrightarrow{F} = \overrightarrow{F}_{drag} + \overrightarrow{F}_{Saffman} + \overrightarrow{F}_{Brownian} = m_p \frac{d\overrightarrow{V}_p}{dt}$$
$$\overrightarrow{F}_{drag} = m_p \frac{3}{8} \frac{\overline{\rho}}{\rho_p} \frac{C_D}{r_p} \left|\overrightarrow{V}_g - \overrightarrow{V}_p\right| \left(\overrightarrow{V}_g - \overrightarrow{V}_p\right)$$
$$\overrightarrow{F}_{Saffman} = m_p \frac{2 K_c (\mu/\rho_g)^{0.5} d_{ij} \rho_g}{\rho_p d_p (d_{lk} d_{kl})^{0.25}} \left(\overrightarrow{V}_g - \overrightarrow{V}_p\right) \tag{13}$$
$$\overrightarrow{F}_{Brownian} = m_p G_0 \sqrt{\frac{\pi S_0}{\Delta t}}$$

where V_g is the gas velocity within the turbulent fluctuation calculated from the gas turbulence model. During the particle tracking procedure, the turbulent dispersion of particles is calculated by integrating the trajectory equations for individual particles, using the instantaneous fluid velocity along the particle path. C_D is the drag force coefficient expressed by [19],

$$C_D = \left(\frac{24}{Re_p} + \frac{6}{1 + \sqrt{Re_p}} + 0.4\right) f_{prop}^{-0.45} f_{Kn}^{0.45} \tag{14}$$

in which the particle Reynolds number is a dimensionless number providing the importance of inertia compared to viscous force and defined as:

$$\text{Re}_p = 2\rho_g r_p \left| \vec{V}_g - \vec{V}_p \right| / \mu \tag{15}$$

f_{prop} represents the effects of variable plasma properties in the boundary layer surrounding the particle, and can be expressed as [20] $f_{prop} = \rho_c \mu_c / \rho_w \mu_w$. f_{Kn} is the factor representing Knudsen effect, which can be expressed by:

$$f_{Kn} = \left[1 + \left(\frac{2-a}{a} \right) \left(\frac{\gamma_w}{1+\gamma_w} \right) \frac{4}{\text{Pr}} Kn \right]^{-1} \tag{16}$$

where a is the thermal accommodation factor, usually with the value of 0.8 [21]; Pr is dimensionless Prandtl number which provides the importance of momentum transfer compared to heat transfer; Kn is the Knudsen number for the importance of small scale effect and defined by the effective mean free pathλ and the droplet diameter, $Kn = \lambda / d_p$, where $\lambda = 2\mu / (\rho_g v_w)$; and v_w is the mean molecular speed that is dependent on the gas temperature near the particle surface T_w, as well as the average molecular weight W of the gas mixture, and can be given as: $v_w = (8RT_w / \pi W)^{1/2}$. For nanoparticles, f_{Kn} is in the range 0.005 to 0.1. For the agglomerates and micro-sized particles, f_{Kn} changes from 0.994 to 0.996 [22].

$K_c = 2.594$ is the constant in the Saffman lift force [23], d_{ij} is the deformation tensor. In the expression of Brownian force, G_0 is a random number between -1 to 1, which is subjected to Gauss distribution. S_0 is the spectral intensity, which can be expressed as $S_0 = (216\mu\sigma_B T_g) / \left(32\pi^2 r_p^5 \rho_p^2 C_c \right)$, Boltzman constant $\sigma_B = 1.38 \times 10^{-23}$ J/K.

Equation (13) is used to depict the particle velocity and trajectory. The local gas conditions around the particle are employed to calculate the particle heating.

Heating and melting of particles: A one-dimensional model was adopted for the particle heating and melting, in which the spherical shape of the particle was assumed. The internal convection within the molten part of the particle was not considered. The temperature distribution inside the particle was described as follows:

$$\rho_p C_p \frac{\partial T_p}{\partial r} = \frac{1}{r^2} \frac{\partial}{\partial r} \left(k_p r^2 \frac{\partial T_p}{\partial r} \right) \tag{17}$$

Zero temperature gradient was assumed in the particle center. The particle surface was subjected to energy conservation law, as [24]:

$$\left. \frac{\partial T_p}{\partial r} \right|_{r=0} = 0, \text{ and } 4\pi r_p^2 \left(k_p \frac{\partial T_p}{\partial r} \right) \bigg|_{r=r_p} = \dot{Q}_{conv} - \dot{Q}_{vap} - \dot{Q}_{rad} \tag{18}$$

where the convection, evaporation latent and radiation heat rates ($\dot{Q}_{conv}$, $\dot{Q}_{vap}$ and $\dot{Q}_{rad}$) are expressed as $4\pi r_p^2 h_f \left(T_f - T_s \right)$, $\dot{m}_v L_v$ and $4\pi r_p^2 \varepsilon_p \sigma_s \left(T_s^4 - T_\infty^4 \right)$, respectively. The film temperature, T_f was defined as $(T_s + T_g)/2$, which is introduced to deal with the steep temperature gradient in the boundary layer around the particle. Only the radiation between the particle surface and the environment was considered in the case of optically thin plasma gas. The heat transfer coefficient, h_f, can be calculated from [25]:

$$\text{Nu} = \frac{2h_f r_p}{k_f} = \left(2.0 + 0.6\text{Re}_p^{1/2}\text{Pr}^{1/3} \right) f_{prop} f_{Kn} f_v \tag{19}$$

where f_v accounts for the effect of mass transfer due to evaporation, which can be found in reference [22]. Additional constraints of energy balance between the heat conduction and latent heat at the melting interface r_m was also considered:

$$\left(k_p \frac{\partial T_p}{\partial r}\right)\bigg|_{r=r_m^-} - \left(k_p \frac{\partial T_p}{\partial r}\right)\bigg|_{r=r_m^+} = L_m \rho_p \frac{dr_m}{dt} \tag{20}$$

3. Experimental and Numerical Setup

The present study was conducted for a direct-current suspension plasma spray system with axial injection of feedstock as shown in Figure 2. The ZrO_2 nanoparticles were suspended in the alcohol solvent, as shown in Figure 4, with solid weight content of 10%. Gas mixtures of argon and hydrogen were ionized in plasma gun to form high temperature and high velocity plasma jet. The droplets and particles were accelerated and heated in the plasma gas. At last, the melting nanoparticles formed coatings on the substrate.

For the plasma gas, the Ar and H_2 gas flow rates are 68 and 12 liters per minute, respectively. The electric power input of the plasma torch gun is 30 kW. The plasma flow field was solved using a three-dimensional cylindrical coordinate system. The radial distance was 6 cm with 57 grid points, the axial distance was 15 cm with 66 grid points, and the circular direction was 2π with 32 grid points. At the axis of the gas field, the symmetrical condition is applied. At the nozzle exit, the velocity and temperature could be expressed by the empirical formulae [26], $v(r) = V_{cl}[1 - (r/R_i)^{1.2}]$, and $T(r) = (T_{cl} - T_w)[1 - (r/R_i)^6] + T_w$, respectively. V_{cl} and T_{cl} are the velocity and temperature on the nozzle exit center line, respectively, which are calculated from the total amount of momentum and thermal energy transferred to the plasma jet. T_w is the wall temperature with the initial value of 300 K, the velocity at the wall boundary is 0. The downstream of the jets flow is open.

The droplets and particles were being tracked in the plasma flow field. The droplets had an initial temperature of 300K, and initial velocity from 50 to 500 m/s. The particle temperature distribution and melting interface were calculated using 50 grid points in spherical coordinates for each solid particle.

4. Validation of Model Predictions

In order to verify the numerical code LAVA-P-3D for plasma spray process, the centerline gas temperature profile were calculated and compared with published numerical results [8] and experimental data [27] in Figure 5. A good agreement between the simulation results and experimental data has been obtained. The deviation between experiment and our modeling result is within 10%, which shows a better accuracy comparing to the FLUENT simulation.

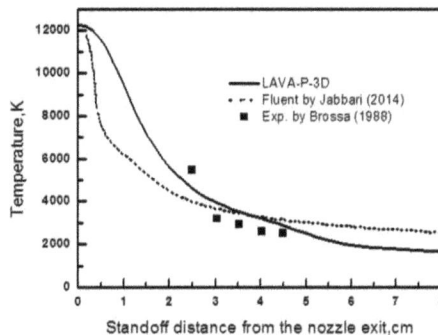

Figure 5. Comparison of gas temperature with published numerical (Reproduced with permission from [8]. Copyright Courtesy of Jabbari, 2014) and experimental data (Reproduced with permission from [27]. Copyright Courtesy of Brossa, 1988).

5. Discussion

5.1. Flow Field of the Plasma Jet

As shown in Figure 6, the hottest plasma zone extends from the torch exit to 1 cm downstream, with temperature as high as 10,000 K. Beyond this core region, the gas temperature cooled down continuously along the spray distance. The gas formed two vortexes near the nozzle exit and another two vortexes in the far field. The vortexes near the exit were formed because of the torch walls besides the exit and the steep shear layer at the outer edge of the jet induced those two vortexes in the far field. We have also noted that vortex location in the far field was not fixed, depending on the specific flow field environment.

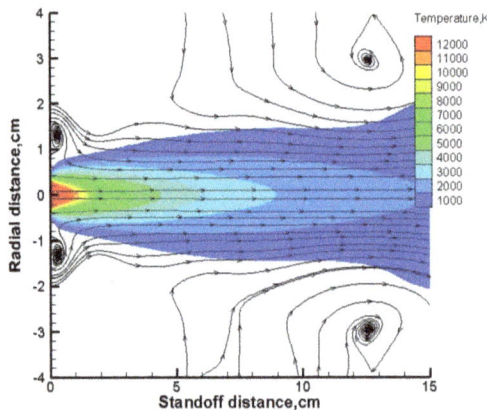

Figure 6. Gas temperature contours in suspension plasma spray.

5.2. Effects of Droplet Diameter on Nanoparticle Release

The sizes of droplet and agglomerate controlled the solvent evaporation and nanoparticle release. Proper atomization parameters contributed to better nanoparticle release. Figure 7 depicts the nanoparticle release position, as well as the solvent evaporation position, from different size of suspension droplets. The solvent evaporation position and nanoparticle release position is the location where solvent evaporation and nanoparticle release take place, respectively, as illustrated in Figure 4. The larger the droplet diameter was, the latter the release position was. This means that the nanoparticle processing in SPS preferred the good atomization and small droplet size.

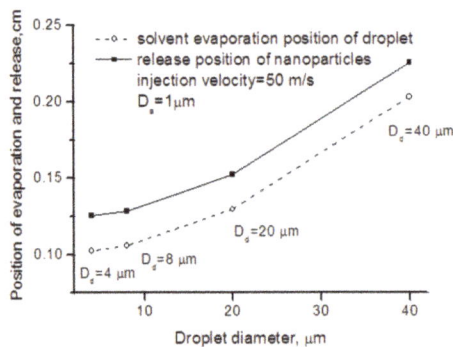

Figure 7. Solvent evaporation position and nanoparticle release position for different droplet diameter.

Figure 8 shows the effects of the agglomerate size on nanoparticle release position, where the droplet diameter was the same. From the picture, we can see that large agglomerate size lengthened the standoff distance between the nanoparticle release and the solvent evaporation, because large agglomerates need more energy to blow the nanoparticles off into the flow field. It should be mentioned that the agglomerate here is not produced from sintering as discussion in the following Sections 5.4 and 5.5. Instead, they are induced from the aggregation effect when preparing the suspension. Thus, it could also be concluded that the nanoparticle processing in SPS preferred good disintegration between the solid and the liquid for suspension preparation.

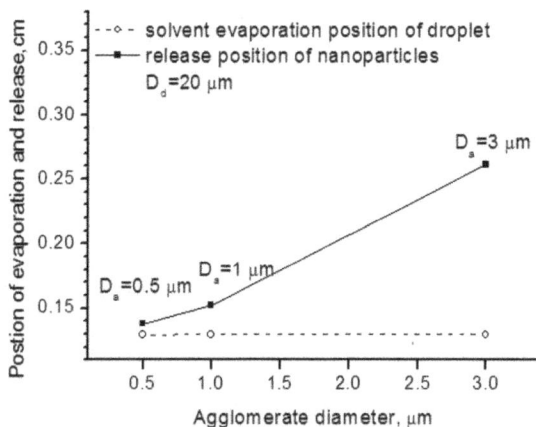

Figure 8. Solvent evaporation position and release position for different agglomerate diameter.

5.3. Effects of Atomization Injection Parameters on Nanoparticle Release

The injection angle and injection velocity are key atomization parameters that affect the nanoparticle release. Injection angle, or cone angle, means the angle between the two outer edges of the spray. Zero cone angle indicates that the spray droplets are injected in the centerline. In experiments, cone angle is larger than zero and here we used cone angles of 10 and 20 degrees. We also defined the nanoparticle release percentage as the ratio of the released nanoparticle mass over the total solid mass.

As shown in Figure 9, when the injection angle became larger, fewer nanoparticles are released into the flow field. The reason is more particles would be distributed at the fringe of the flow field and could not be well heated up if the injection angle is larger. The particle injection velocity also affected the release percentage of the nanoparticles. Two cases with different initial particle injection velocity of 50 m/s and 500 m/s are simulated. Results in Figure 10 showed that, when the injection velocity increasing, the release percentage of nanoparticles became smaller. This is because the high-velocity particles have short flight and heating time in the high-temperature flow field. In addition, the results showed that the nanoparticle release occurred early before the standoff distance of 0.4 cm, since the solvent evaporation point is pretty low at about 350 K while the plasma gas in this region is much higher.

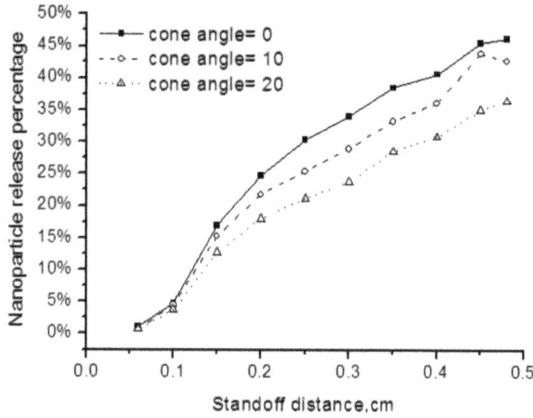

Figure 9. Effects of injection angle on the nanoparticle release percentage.

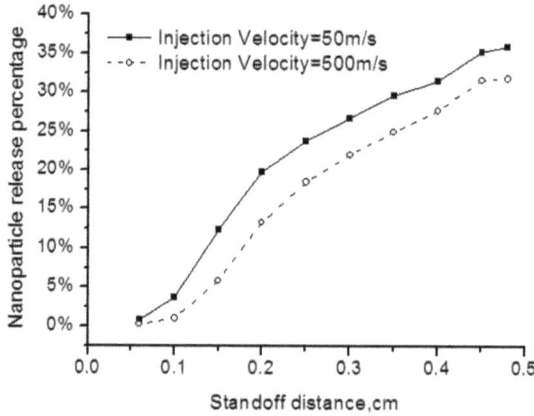

Figure 10. Effects of injection velocity on the nanoparticle release percentage.

5.4. Velocity, Temperature and Melting of Nanoparticles and Agglomerates

As shown in Figure 4, many suspension droplets would be fragmentized into small pieces by aerodynamics force, which would be released into nano-sized particles after solvent evaporation. However, some of the suspension droplets might not experience enough aerodynamics force to further breakup the droplets (see Figure 3), which would be sintered into a micro-sized solid agglomerate after solvent evaporation. Those released nano-sized particles and sintered micro-sized agglomerates have different heating and accelerating processes.

Figure 11 represents the particle velocity for nanoparticle and agglomerate with different diameter. In this study, the released nanoparticles have diameter ranging from 10 to 100 nm. Results showed that the small inertia of nanoparticles improved the acceleration process, especially in the hottest plasma zone within 1 cm from the exit where the plasma velocity is also very high. However, the nanoparticles decelerated rapidly behind this zone, for the reason of slowing plasma gas. This means that the nanoparticles show good tracking characteristics, and their impact velocity before coating mainly depends on the local gas velocity. For the micro-sized agglomerates, their acceleration processes were much less than that of nanoparticles due to their large inertia. These micron particles usually keep accelerating during their flight. Therefore, their impact velocity mainly depends on the speeding rate, and the relatively small micro-sized agglomerates have high impact velocity.

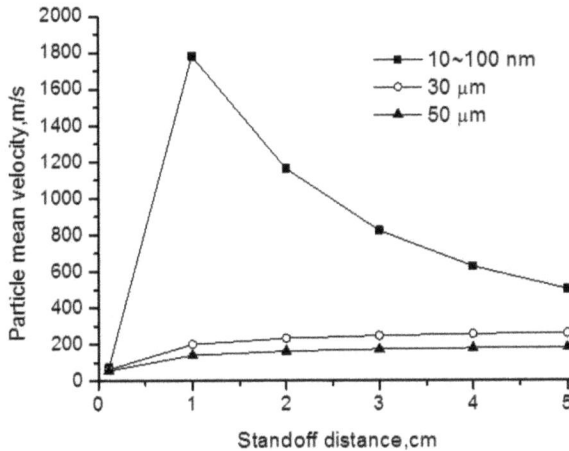

Figure 11. Velocity of nanoparticles and agglomerates.

The particle temperature and melting status are shown in Figure 12 for nanoparticles and agglomerates with different diameter. From the picture, we can see that the nanoparticles could be easily melted when reaching melting temperature at the standoff distance 0.6 cm. After being totally melted, the nanoparticles continued heating, and the temperature could reach the solid evaporation point as high as 5000 K. For the agglomerates, the heating rate was much less than that of the nanoparticles. The larger the particle was, the lower the temperature reached. For this reason, some agglomerates could be melted while others could not. Results showed that the 30 µm agglomerate was totally melted at the distance of 1.1 cm and the 50 µm agglomerate was melted at 2.9 cm, but the 90 µm agglomerate never reached the melting temperature. The larger the particles were, the more heat needed to absorb from the flow field to reach melting status. Therefore, the melting position was delayed more to the rear of flow field if increasing the particle diameter. This also indicated that the solvent evaporation was not a key factor for agglomerate temperature and melting, since evaporation had been done early before standoff distance about 0.4 cm.

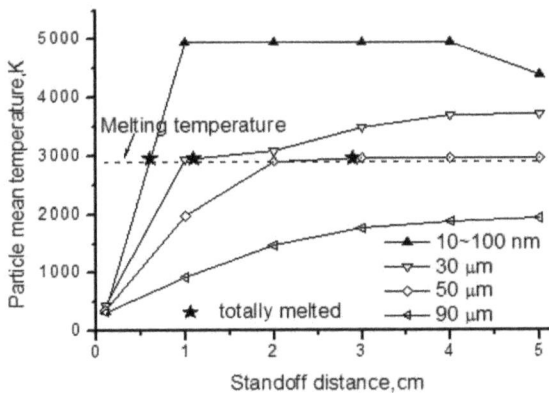

Figure 12. Temperature of nanoparticles and agglomerates.

5.5. Critical Agglomerate and Droplet Size

Particle diameter was important to their melting status before coating, especially for agglomerates. We define the melting percentage for nanoparticles and agglomerates, as the ratio of the melted mass over the initial solid mass during their flight.

Table 1 lists the maximum of melting percentage for nanoparticles and agglomerates with different size, as well as the standoff position where this maximal melting finished. Results showed that all the nanoparticles could be easily melted once they were released and heated up to melting temperature. For micro-sized agglomerates, things were different in two aspects. Firstly, their melting percentage decreased continuously with the particle size increased. Secondly, their flight distance before being mostly melted was much longer comparing to the nanoparticles. One reason is that the heat transfer efficiency of agglomerates is lower than that of the nanoparticles. Another reason is that the agglomerates have large latent heat at the melting interface and heat conduction needs a long flight time to balance it. Poorly melted particles would reduce the coating quality and their existence should be avoided. For this reason, the agglomerate size should be limited to a critical number, as well as the droplet size.

Table 1. Melting percentage and position for nanoparticles and agglomerates.

Particle Diameter	Maximum of Melting Percentage	Mostly Melted Position
10–100 nm	100%	6 mm
30 μm	100%	11 mm
40 μm	100%	15 mm
50 μm	100%	29 mm
60 μm	93.26%	38 mm
70 μm	76.40%	62 mm
80 μm	2.25%	74 mm
90 μm	0	None

Our calculation showed the critical agglomerate diameter being totally melted is 50 μm for current SPS conditions. The agglomerates smaller than this size could be totally melted and those larger than it could only be partially melted or not melted at all. Our previous results showed that the solvent evaporation was not a key factor for agglomerate temperature and melting. Thus, we could assume that a suspension droplet with diameter of $D_{d,crit}$ and solid weight content of wt, experienced no secondary atomization and was sintered to one agglomerate with diameter of $D_{a,crit}$, *i.e.*, 50 μm in this case. Then, the critical suspension droplet size, $D_{d,crit}$ could be estimated as:

$$D_{d,crit} = D_{a,crit} (wt)^{-1/3} \tag{21}$$

For the typical weight content of 10%–25%, the critical suspension droplet size could be calculated as 80–100 μm for current suspension plasma spray. Lower weight content had larger critical droplet size. Such estimation agrees fairly well with what Fauchais reported in 2008, that the droplet diameter should be limited below 90 μm for better particle melting [28].

6. Conclusions

In our present work, a comprehensive model has been developed to simulate the evolution of suspension droplet, nano-sized particles, as well as micro-sized agglomerates in the high-temperature plasma jet. This model simultaneously studied the dynamic, thermal and phase change behaviors of nano-sized or micro-sized particles. To validate this model, we compared our numerical results with other group's experiments as well as simulation work, proving its accuracy with deviation less than 10%. Using this model, effects of atomization parameters on the nanoparticle processing have been thoroughly studied. Proper conditions are concluded to ensure nanoparticle release and agglomerate melting during suspension plasma spray.

Firstly, the effects of atomization parameters on the nanoparticle release, including droplet diameter, injection angle and velocity, were investigated. The study reveals that the nanoparticle processing in SPS preferred small droplets with better atomization and less aggregation from suspension preparation. The injection angle and velocity influenced the nanoparticle release percentage. Small angle and low initial velocity might have more nanoparticles released.

Besides, the melting percentage of nanoparticles and agglomerates were studied, and the critical droplet diameter to ensure solid melting was drawn. Results showed that for better solid melting, the suspension droplet size should be limited to a critical droplet diameter, which was inversely proportional to the cubic root of weight content of the given critical agglomerate diameter of being totally melted. Of course, this critical agglomerate diameter would depend on the plasma condition, the particle material and the substrate distance, which could be further studied in the future.

Acknowledgments: The work was supported by the National Natural Science Foundation of China (No. 11472245), and the Fundamental Research Funds for the Central Universities (No. 2012FZA4027).

Author Contributions: Hong-bing Xiong designed the study. Cheng-yu Zhang and Kai Zhang wrote the paper. Xue-ming Shao edited the manuscript.

Conflicts of Interest: The authors declare no conflict of interest.

Nomemclature

C_D	drag coefficient
C_p	specific heat, $J \cdot kg^{-1} \cdot K^{-1}$
D_d	droplet diameter, µm
D_a	agglomerate diameter, µm
d_{ij}	deformation tensor
d_p	particle diameter, m
f_{Kn}	factor ofKnudsen effect
f_{prop}	factor to account property variation
f_v	factor formass transfer
G_0	zero-mean, unit-variance independent Gauss random number
h	heat transfer coefficient, $W \cdot m^{-2} \cdot K^{-1}$ or the half-thickness of sheath
k	thermal conductivity, $W \cdot m^{-1} \cdot K^{-1}$ or wave number
K_c	constant coefficient in Saffman lift force
Kn	Knudsen number
L_m	latent heat of fusion, $J \cdot kg^{-1}$
L_v	latent heat of vaporization, $J \cdot kg^{-1}$
m_p	particle mass, kg
$\dot{m}_v$	vaporization rate, $kg \cdot s^{-1}$
Nu	Nusselt number
p	pressure, Pa
Pr	Prandtl number, $Pr = \mu C_p k^{-1}$
Q	heat flux, $W \cdot m^{-2}$ or the gas/liquid density ratio
R	gas constant, $J \cdot mol^{-1} \cdot K^{-1}$
r	radial coordinate for the particle, m
r_m^-	inner radius of melting interface, m
r_m^+	outer radius of melting interface, m
r_m	position of particle melting interface, m
r_p	particle radius, m
S_0	spectral intensity

SMD	Sauter mean diameter, μm
T	temperature, K
t	time, s
T_∞	temperature outside boundary layer, K
$T_{d,0}$	droplet initial temperature, K
$T_{m,d}$	droplet melting point, K
T_s	particle surface temperature, K
T_w	gas temperature near particle surface, K
v_g	gas velocity, m·s^{-1}
V	velocity, m·s^{-1}
v_w	mean molecular speed, m·s^{-1}
W	molecular weight of thegas mixture, kg·mol^{-1}
Weber	Weber number, dimensionless
wt	solid weight content in suspension, dimensionless
y	deformation parameter, dimensionless
σ_s	Stefan-Boltzmann constant

Greek symbols

ρ	density, kg·m^{-3}
μ	viscosity, kg·s^{-1}·m^{-1}
γ_w	specific heat ratio of gas, dimensionless
α	weight fraction or void fraction
ε	surface emissivity coefficient
δ	the thickness of annular liquid sheath
σ	surface tension
ω	the growth rate of surface disturbances

Subscript

d	suspension droplet embedded with nanoparticles
f	film around the particle
g	plasma gas
p	solid nanoparticles or agglomerates
sl	solvent

References

1. Fauchais, P.; Montavon, G.; Lima, R.S.; Marple, B.R. Engineering a new class of thermal spray nano-based microstructures from agglomerated nanostructured particles, suspensions and solutions: An invited review. *J. Phys. D* **2011**, *44*. [CrossRef]
2. Ye, F.; Ohmori, A.; Li, C. New approach to enhance the photocatalytic activity of plasma sprayed TiO$_2$ coatings using p-n junctions. *Surf. Coat. Technol.* **2004**, *184*, 233–238. [CrossRef]
3. Meillot, E.; Vincent, S.; Caruyer, C.; Damiani, D.; Caltagirone, J.P. Modelling the interactions between a thermal plasma flow and a continuous liquid jet in a suspension spraying process. *J. Phys. D* **2013**, *46*. [CrossRef]
4. Marchand, C.; Chazelas, C.; Mariaux, G.; Vardelle, A. Liquid precursor plasma spraying: Modeling the interactions between the transient plasma jet and the droplets. *J. Ther. Spray Technol.* **2007**, *16*, 705–712. [CrossRef]
5. Huang, P.C.; Heberlein, J.; Pfender, E. Particle behavior in a two-fluid turbulent plasma jet. *Surf. Coat. Technol.* **1995**, *73*, 142–151. [CrossRef]
6. Fazilleau, J.; Delbos, C.; Rat, V.; Coudert, J.F.; Fauchais, P.; Pateyron, B. Phenomena involved in suspension plasma spraying part 1: Suspension injection and behavior. *Plasma Chem. Plasma Process.* **2006**, *26*, 371–391. [CrossRef]

7.   Ozturk, A.; Cetegen, B.M. Modeling of axially and transversely injected precursor droplets into a plasma environment. *Int. J. Heat Mass Transf.* **2005**, *48*, 4367–4383. [CrossRef]
8.   Jabbari, F.; Jadidi, M.; Wuthrich, R.; Dolatabadi, A. A Numerical Study of Suspension Injection in Plasma-Spraying Process. *J. Ther. Spray Technol.* **2014**, *23*, 3–13. [CrossRef]
9.   Lund, M.T.; Sojka, P.E.; Lefebvre, A.H.; Gosselin, P.G. Effervescent atomization at low mass flow rates. Part I: The influence of surface tension. *Atom. Sprays* **1993**, *3*, 77–89. [CrossRef]
10. Ishii, M. *One-Dimensional Drift-Flux Model and Constitutive Equations for Relative Motion between Phases in Various Two-Phase Flow Regimes*; Argonne National Laboratory: Argonne, IL, USA, 1977; pp. 47–77.
11. Weber, C. Disintegration of liquid jets. *Z. Angew. Math. Mech.* **1931**, *11*, 136–159. [CrossRef]
12. Senecal, P.K.; Schmidt, D.P.; Nouar, I.; Rutland, C.J.; Reitz, R.D.; Corradini, M.L. Modeling high-speed viscous liquid sheet atomization. *Int. J. Mult. Flow* **1999**, *25*, 1073–1097. [CrossRef]
13. Xiong, H.B.; Zheng, L.L.; Sampath, S.; Williamson, R.L.; Fincke, J.R. Three-dimensional simulation of plasma spray: Effects of carrier gas flow and particle injection on plasma jet and entrained particle behavior. *Int. J. Heat Mass Transf.* **2004**, *47*, 5189–5200. [CrossRef]
14. Tanner, F.X. Development and validation of a cascade atomization and drop breakup model for high-velocity dense sprays. *Atom. Sprays* **2004**, *14*, 211–242. [CrossRef]
15. O'Rourke, P.J.; Amsden, A.A. *The TAB Method for Numerical Calculation of Spray Droplet Breakup*; No. 872089, SAE Technical Paper; SAE International: Warrendale, PA, USA, 1987.
16. Reitz, R.D. Modeling atomization processes in high-pressure vaporizing sprays. *Atom. Spray Technol.* **1987**, *3*, 309–337.
17. Delbos, C.; Fazilleau, J.; Rat, V.; Coudert, J.F.; Fauchais, P.; Pateyron, B. Phenomena involved in suspension plasma spraying Part 2: Zirconia particle treatment and coating formation. *Plasma Chem. Plasma Process.* **2006**, *26*, 393–414. [CrossRef]
18. Fauchais, P.; Montavon, G. Latest developments in suspension and liquid precursor thermal spraying. *J. Ther. Spray Technol.* **2010**, *19*, 226–239. [CrossRef]
19. Chen, X.; Pfender, E. Behavior of small particles in a thermal plasma flow. *Plasma Chem. Plasma Process.* **1983**, *3*, 351–366. [CrossRef]
20. Lee, Y.C.; Hsu, K.C.; Pfender, E. Modeling of particles injected into a DC plasma jet. In Proceedings of the 5th International Symposium on Plasma Chemistry, Edinburgh, UK, 10–14 August 1981; Volume 2, p. 795.
21. Chen, X.; Pfender, E. Effect of the Knudsen number on heat transfer to a particle immersed into a thermal plasma. *Plasma Chem. Plasma Process.* **1983**, *3*, 97–113. [CrossRef]
22. Xiong, H.B.; Lin, J.Z. Nanoparticles modeling in axially injection suspension plasma spray of zirconia and alumina ceramics. *J. Ther. Spray Technol.* **2009**, *18*, 887–895. [CrossRef]
23. Saffman, P.G.T. The lift on a small sphere in a slow shear flow. *J. Fluid Mech.* **1965**, *22*, 385–400. [CrossRef]
24. Wan, Y.P.; Prasad, V.; Wang, G.; Sampath, S.; Fincke, J.R. Model and powder particle heating, melting, resolidification, and evaporation in plasma spraying processes. *J. Heat Transf.* **1999**, *121*, 691–699. [CrossRef]
25. Chen, X.; Pfender, E. Heat transfer to a single particle exposed to a thermal plasma. *Plasma Chem. Plasma Process.* **1982**, *2*, 185–212. [CrossRef]
26. Ramshaw, J.D.; Chang, C.H. Computational fluid dynamics modeling of multicomponent thermal plasmas. *Plasma Chem. Plasma Process.* **1992**, *12*, 299–325. [CrossRef]
27. Brossa, M.; Pfender, E. Probe measurements in thermal plasma jets. *Plasma Chem. Plasma Process.* **1988**, *8*, 75–90. [CrossRef]
28. Fauchais, P.; Etchart-Salas, R.; Rat, V.; Coudert, J.F.; Caron, N.; Wittmann-Ténèze, K. Parameters controlling liquid plasma spraying: Solutions, sols, or suspensions. *J. Ther. Spray Technol.* **2008**, *17*, 31–59. [CrossRef]

nanomaterials

MDPI

Article

Post-Plasma SiO$_x$ Coatings of Metal and Metal Oxide Nanoparticles for Enhanced Thermal Stability and Tunable Photoactivity Applications

Patrick Post [1], Nicolas Jidenko [2], Alfred P. Weber [1,*] and Jean-Pascal Borra [2]

[1] Institute of Particle Technology, Clausthal University of Technology, Leibnizstrasse 19,
    38678 Clausthal-Zellerfeld, Germany; patrick.post@tu-clausthal.de
[2] Lab of Phys Gaz and Plasmas, CNRS, Univ. Paris Sud, CentraleSupelec, Université Paris-Saclay,
    F-91405 Orsay, France; nicolas.jidenko@u-psud.fr (N.J.); jean-pascal.borra@u-psud.fr (J.-P.B.)
*   Correspondence: weber@mvt.tu-clausthal.de; Tel.: +49-532-372-2309

Academic Editors: Krasimir Vasilev and Melanie Ramiasa
Received: 9 March 2016; Accepted: 9 May 2016; Published: 13 May 2016

Abstract: The plasma-based aerosol process developed for the direct coating of particles in gases with silicon oxide in a continuous chemical vapor deposition (CVD) process is presented. It is shown that non-thermal plasma filaments induced in a dielectric barrier discharge (DBD) at atmospheric pressure trigger post-DBD gas phase reactions. DBD operating conditions are first scanned to produce ozone and dinitrogen pentoxide. In the selected conditions, these plasma species react with gaseous tetraethyl orthosilicate (TEOS) precursor downstream of the DBD. The gaseous intermediates then condense on the surface of nanoparticles and self-reactions lead to homogeneous solid SiO$_x$ coatings, with thickness from nanometer to micrometer. This confirms the interest of post-DBD injection of the organo-silicon precursor to achieve stable production of actives species with subsequent controlled thickness of SiO$_x$ coatings. SiO$_x$ coatings of spherical and agglomerated metal and metal oxide nanoparticles (Pt, CuO, TiO$_2$) are achieved. In the selected DBD operating conditions, the thickness of homogeneous nanometer sized coatings of spherical nanoparticles depends on the reaction duration and on the precursor concentration. For agglomerates, operating conditions can be tuned to cover preferentially the interparticle contact zones between primary particles, shifting the sintering of platinum agglomerates to much higher temperatures than the usual sintering temperature. Potential applications for enhanced thermal stability and tunable photoactivity of coated agglomerates are presented.

Keywords: particle coating; silicon oxide; dielectric barrier discharge; gas phase; continuous process; TEOS; CVD

1. Introduction

The coating of nanoparticles is an essential step to improve the particle thermostability [1], to create a protective layer between particle and environment [2] or to enhance particle dispersion in liquids. Silicon oxide is one of the preferred nontoxic and almost inert coating materials.

Many coating techniques use liquid phase reactions with additional steps such as washing, drying and separation [3]. Moreover, impurities from solvents remain in the final solid coatings. Therefore, gas phase coating techniques are generally preferred [4] (e.g., for atomic layer deposition [5], for flames [6,7] and for plasma enhanced chemical vapor deposition (CVD) [8]).

Most publications on SiO$_x$ coatings deal with the coating of flat surfaces, but few groups used it to coat nanoparticles directly in the gas phase. Homogeneous coatings can be performed in non-continuous devices such as fluidized bed reactors [9]. In flame synthesis, direct coating of gas born nanoparticles such as TiO$_2$, may be performed in a continuous one-step process.

A more versatile coating method is the CVD that can be induced by photochemistry [10] and plasmas. In dielectric barrier discharges (DBD), plasma filaments produce reactive species (electrons, photons, radicals, excited species and more stable ozone and nitrogen oxides in air), triggering the polymerization of organic and organo-metal precursors. For the coating of substrates, the precursors can be injected in the DBD or in post-DBD for SiO_x coatings [8,11], as well as for functional polymer coatings [12,13]. For the coating of supported particles by plasma, Mori *et al.* [14], Kogoma *et al.* [15] and Brüser *et al.* [16] used batch processes.

The direct SiO_x coating of suspended nanoparticles in gases using atmospheric pressure DBD is an emerging field [13]. Nessim *et al.* used HMDSO (hexamethyldisiloxane) and TEOS for the coating of particles either directly in a plasma as Vons *et al.* [17], or in post-DBD [18]. As for flat surface coating, the precursor injection in the DBD leads to a loss of functionality of the gaseous precursor reacting with plasma species in the gap, to particles electro-collection on surfaces in the gap and to discharge destabilization that hampers the economics of the process.

Here, post-DBD injection of the organo-silicon precursor is tested to avoid electrode coating so as to achieve stable DBD production of active species with subsequent controlled thickness of SiO_x coatings of nanoparticles *versus* tetraethyl orthosilicate (TEOS) concentration and reaction duration. The setup and operating conditions are presented. Preliminary tests confirm that plasma species trigger the post-DBD conversion of TEOS into solid SiO_x coatings of nanoparticles. DBD operating conditions are then scanned to produce ozone and dinitrogen pentoxide. In the selected conditions, subsequent sections present the coating of spherical particles and of agglomerates separately *versus* reaction duration and concentrations of reactants (plasma species and TEOS). Finally, properties of coated agglomerates for thermal stabilization of catalyst particles and for photoactivity control are depicted.

2. Experimental Section

The experimental setup is shown in Figure 1. The first step is the production of nanoparticles with defined size, concentration and morphology (spheres or agglomerates). Then, a controlled amount of precursor is added. The third step is the post-DBD injection of this TEOS/nanoparticles aerosol, mixed with plasma species. Finally, post-DBD reaction times are tuned in different volumes.

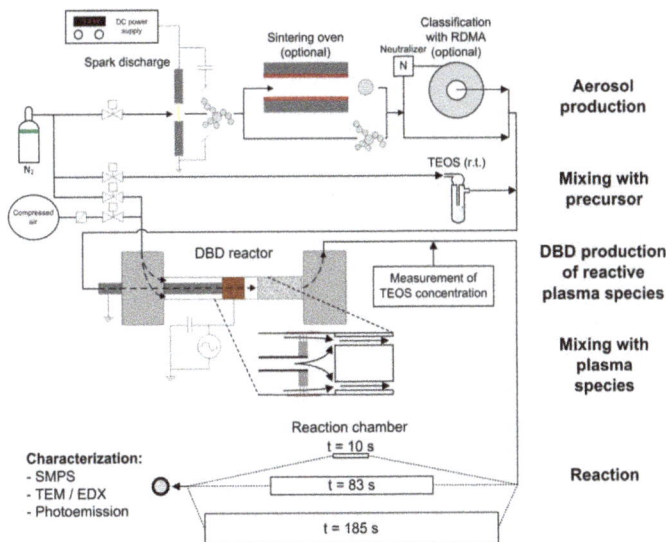

Figure 1. Experimental setup with the different process steps.

2.1. Aerosol Production and Mixing with Precursor

The metal core particles were produced in a spark discharge generator (SDG) without any side products in the gas stream [19]. Two electrodes of the same material with a 5 mm gap were polarized with a high voltage DC power supply. Then, repetitive sparks develop between the electrodes leading to vaporization and nucleation into nanoparticles [19]. The preferred material was platinum for its high TEM contrast and applications, e.g., in catalysis. Due to the high number concentrations of small primary particles large fractal agglomerates formed rapidly behind the SDG in 1 L/min N_2 (standard liter per minute, nitrogen 5.0). Figure 2 (left) shows such agglomerates, converted into spherical particles (right) by sintering in a tube furnace at 1000 °C. Pt, Cu, Ti, Ni and Fe have been used in the spark generator to produce similar nanoparticles, agglomerated or spherical after sintering, which could be size selected with a Radial Differential Mobility Analyzer (RDMA) downstream of an X-ray source to establish a known charge distribution.

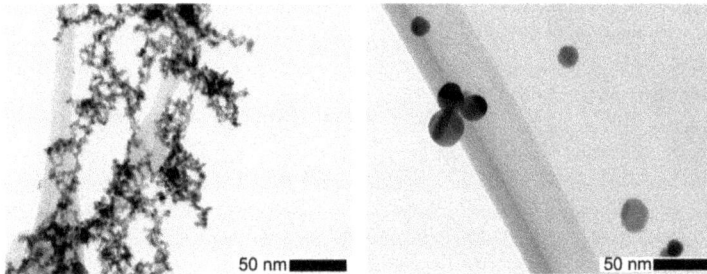

Figure 2. Effect of sintering on the morphology of platinum nanoparticles from the spark generator: (**left**) agglomerates downstream of the spark before sintering; and (**right**) spherical particles after sintering.

Among the organo-silicon precursors, tetraethyl orthosilicate (TEOS, $Si(OC_2H_5)_4$) is one of the most prominent. Due to its non-hazardous character, TEOS is easy to handle. Moreover, its relatively low vapor pressure facilitates the addition of small amounts of gaseous precursor and it reacts more slowly compared to other precursors and is therefore suitable to achieve nano-sized SiO_x coatings.

The mixing of the TEOS with the particles is done in a simple T-piece prior to post-DBD injection. A second flow of nitrogen below 30 mL/min is used to transport the precursor vapor from a bottle. The nitrogen does not bubble through the precursor but flows over the surface of the liquid to prevent the formation and transport of droplets. In the bottle kept at room temperature (24 °C), the distance between the gas inlet and the surface of the initially liquid TEOS is kept constant at 5 mm.

The precursor concentration was measured by Fourier transform infrared spectroscopy (FTIR, Tensor 27, Bruker Optik GmbH, Ettlingen, Germany) on the Si–O absorbance near 1100 cm^{-1}, directly after the mixing of all gas flows downstream of the DBD reactor (Figure 1), *i.e.*, at the entrance of the post-DBD reaction chamber. It can be tuned from 0.3 to 4.1 ppmv (below 0.2% of TEOS saturation vapor pressure of 242 Pa at 24 °C), without affecting significantly the total flow of 3 L/min (plus a maximum of 30 mL/min of the N_2-TEOS mixture).

2.2. DBD Arrangement and Operating Conditions

The DBD is made of two stainless steel parts that hold a quartz glass tube with an inner diameter of 13 mm and a thickness of 1.5 mm serving as dielectric (see inset of Figure 1). On the outside of the glass tube, an electrode made of copper adhesive tape is connected to the high voltage AC power supply. A stainless steel disc (diameter: 12 mm, length: 2 mm) is fixed on the grounded central stainless steel tube in front of the polarized external electrode. Plasma filaments then occur in the 0.5 mm ring-shaped gap between the metal disc and the dielectric, in a so-called monoDBD [20].

To limit the electrode temperature below 100 °C, the generator frequency was fixed to 41 kHz. The voltage was measured with a HV probe. The current pulses related to each plasma filament were recorded via a 50 Ω input resistor to evaluate the charge and the energy per filament as well as the power, calculated, with n the number of periods and T the period duration, according to:

$$P = \frac{1}{n \cdot T} \cdot \int_{nT} u(t) \cdot i(t) \, dt \tag{1}$$

2.3. Mixing with Plasma Species and Reaction

The particles and the precursor flow through a stainless steel tube (4.57 mm inner diameter) without contact with the plasma. The combined aerosol flow from the spark generator and the TEOS bottle was injected into the inner tube at 1 L/min N_2 varying only slightly with the small amount of precursor gas flow. A 1:1 mixture of nitrogen and filtered air was injected at 2 L/min into the DBD gap to transport plasma species downstream of the DBD up to the post-DBD mixing with the aerosol/precursor mixture. The total flow of about 3 L/min was injected in the post-DBD reaction volume. To enhance the mixing of plasma species and of the particle/TEOS aerosol flow, a static mixer was placed about 8 mm downstream of the DBD so that there is no reflux of TEOS into the discharge zone. As shown in Figure 1, it deflects the inner flow towards smaller tubes near the outer diameter of the mixer, leading to turbulent mixing of the two gas flows.

The influence of the reaction duration was tested with different tubes as reaction chambers with laminar flow leading to defined transit time distributions. Unless stated otherwise, the selected conditions defined from the preliminary results detailed in Section 3.2, are specified in Table 1.

Table 1. Summary of the standard experimental conditions.

Description	Abbr.	Value
Volume flow through discharge	Q_{Plasma}	2 L/min
Volume flow in particle generation step	$Q_{Particles}$	1 L/min
TEOS concentration	c_{TEOS}	0.7 ppmv
Oxygen concentration in discharge	c_{Oxygen}	10.5%
Applied peak-to-peak voltage	U_{PP}	10 kV
Applied plasma power (Figure 3)	P_{Plasma}	4 W
Reaction time before sampling	$t_{Reaction}$	83 s

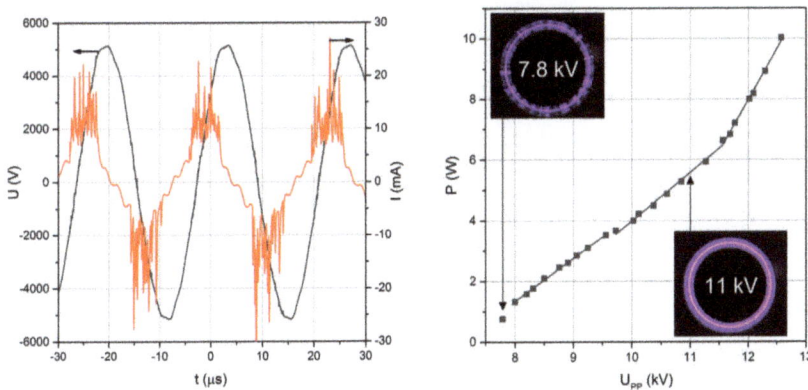

Figure 3. Temporal evolution of voltage and dielectric barrier discharge (DBD) current (**left**) and DBD power *versus* peak-to-peak voltage characteristic of the DBD with insets of violet light emitted from plasmas in air (**right**).

2.4. Aerosol Characterization

Downstream of the reaction chamber, the coated particles were collected at 0.2 L/min in a bypass across a lacey transmission electron microscopy (TEM) grid for 2 min. As a consequence, the particles analyzed with the TEM refer to different ages since condensation and heterogeneous reactions still happen during the 2 min collection. However, for the presentation of the experimental data, the beginning of the sampling was chosen as time reference. Then, the TEM grid was removed and stored in ambient air for at least 20 min before being transferred under ultra-high vacuum into the TEM. Coupled with the energy-dispersive X-ray spectroscopy (EDX) sensor, size, coating thickness and atomic composition of the coated particles were characterized.

On-line measurements of the particle number size distributions were obtained with a scanning mobility particles sizer (SMPS Grimm 5.403, GRIMM Aerosol Technik GmbH & Co. KG, Ainring, Germany). Assuming spherical particles, the coating thickness is defined as half the difference of the mobility equivalent mode diameters of non-coated and coated particles.

The coated surface of particles was estimated by aerosol photoemission measurements (APE) [21]. Downstream of an electrostatic precipitator, the neutral aerosol exposed to UV is charged positively due to electron emission from metal nanoparticles. The aerosol charge, measured with a Faraday Cup Electrometer after an ion trap is proportional to the uncoated active surface area.

3. Results and Discussion

3.1. DBD Electrical Characterisation and Post-DBD Ozone and NO_x

As shown in Figure 3 (left), both voltage and current are sinusoidal with a phase shift between them. The real discharge current with pulses related to each plasma filament is superposed to this sinusoidal capacitive current. From the instantaneous voltage and current, the discharge power was calculated as detailed above and depicted in Figure 3 (right) *vs.* applied voltage.

At increasing voltages, the plasma filaments originate from the edges of the inner electrode until they cover the whole surface of the smaller electrode for higher applied voltages (see insets of Figure 3). As expected for such asymmetrical mono-DBD [20], negative current pulses for filaments developing from the metal electrode to the dielectric surface are slightly larger than positive pulses related to the same streamer-like plasma filament, then directed towards the central metal electrode.

Additional post-DBD FTIR analysis of gaseous products without any TEOS reveal ozone and NO_x, respectively, from 0 to 301 ppmv for ozone and from 35 to 114 ppmv for NO_2, N_2O, N_2O_5 and HNO_x, without NO, as depicted in Table 2 for different voltages. The species were quantified from the adsorption signal at the corresponding wavelengths of each species: O_3: 1050 cm^{-1}, NO_2: 1630 cm^{-1}, N_2O: 2340 cm^{-1}, N_2O_5: 1240 cm^{-1}, NO: 1890 cm^{-1} and HNO_x: 1330 cm^{-1}. These species were expected from a DBD fed with a 10% O_2/90% N_2 mixture, since these DBD were used as ozonizers.

Table 2. Electrode and reaction temperatures, and ozone and NO_x concentrations (in ppmv) for three DBD voltages. * HNO_x could not be quantified due to significant interference.

$U_{DBD, PP}$ (kV)	$T_{electrode}/T_{reaction}$ (°C)	O_3	NO_2	N_2O	N_2O_5	NO	HNO_x	Total NO_x
8	33/20	267 ± 9	23 ± 8	2 ± 2	10 ± 0	0	*	35
10	46/35	301 ± 3	20 ± 0	20 ± 4	24 ± 0	0	*	64
12	120/<70	0	76 ± 5	38 ± 4	0	0	*	114

Finally, it can be noted from electrode temperatures depicted in Table 2 that up to 10 kV the gas is not heated above 100 °C in the gap of the DBD since ozone destruction would be faster than its production. On the contrary, at 12 kV, ozone and dinitrogen pentoxides are converted into other NO_x, as expected for higher gas temperatures arising from higher electrode temperatures.

In the selected arrangement, a stable DBD with constant energy and number of pulses per cycle leads to constant related post-DBD concentrations of plasma species at the entrance of the reaction volume. These concentrations can be tuned with DBD operating conditions with ozone and N_2O_5 up to 10 kV or only with other nitrogen oxides at 12 kV, for reaction temperatures near ambient temperatures up to 35 °C for 8 and 10 kV and below 70 °C at 12 kV, respectively.

3.2. Proof of Concept and Ranges of TEOS Concentration and Reaction Duration for Nanoparticle Coating

It is first confirmed that post-DBD gas phase reactions of ozone and dinitrogen pentoxide plasma products with gaseous TEOS precursor lead to final homogeneous SiO_x coatings.

This can be stated from EDX analysis presented below in Table 3 and from Figure 4 (left). The size distributions of sintered particles do not evolve by mixing with TEOS precursor unless plasma filaments develop in the DBD. Then, a significant increase in diameter is found when the DBD is fed with a 10% O_2/90% N_2 mixture (Figure 4, left), while in pure nitrogen, *i.e.*, without oxygen injected either in the DBD or in post-DBD, some holey heterogeneous coatings are formed. As sub-second transit time from the DBD gap (where plasma filaments produce electrons, photons, radicals, excited species) to the entrance of the reaction volume is longer than the times of life of these plasma species (e.g., for atomic oxygen—10^{-8} s for O^1d, 10^{-8} to 10^{-7} s for O^3p- and metastable $O_2{}^1\Delta g$ and vibrationally excited N_2, up to 10^{-3} s, in air [22]), only stable species such as ozone and nitrogen oxides are analyzed in post-DBD at the entrance of the reaction volume and could react with TEOS.

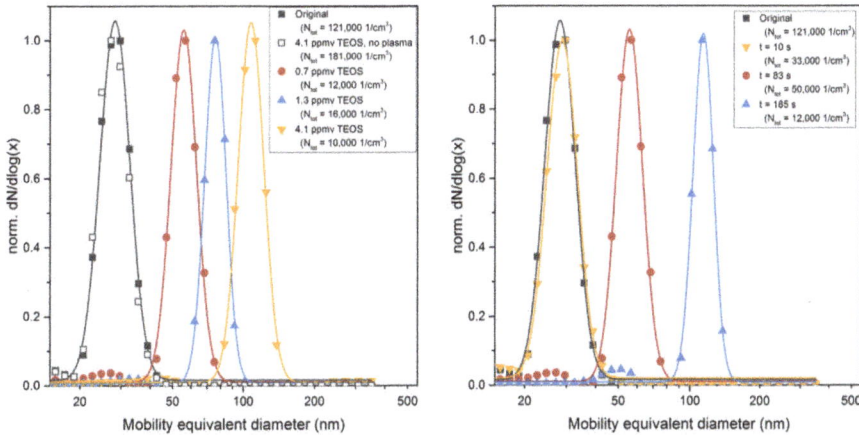

Figure 4. Mobility equivalent diameter measured with scanning mobility particles sizer (SMPS): (**left**) as a function of the tetraethyl orthosilicate (TEOS) concentration at constant reaction time $t_{Reaction}$ = 83 s; and (**right**) as a function of reaction time at constant TEOS concentration of 0.7 ppmv (the total number concentration N_{tot} is given in the legend).

Table 3. Elemental composition from the quantitative energy-dispersive X-ray spectroscopy (EDX) analysis as a mean of several measurements on the samples for two different reaction times.

Element	$t_{Reaction}$ = 83 s	$t_{Reaction}$ = 185 s
Si	5 at.%	23 at.%
O	13 at.%	48 at.%
C	30 at.%	23 at.%
Pt (core)	53 at.%	6 at.%

Actually, Okuyama *et al.* [23] and Fujino *et al.* [24] studied the kinetics of TEOS and ozone reactions, which was later modeled by Romet *et al.* [25]. Atomic oxygen produced by ozone decomposition is suspected to be the main starter of the reaction [25]. One critical step is the gas phase conversion of TEOS into triethoxysilanol ($Si(OC_2H_5)_3OH$), then condensed on the surface and finally converted into solid SiO_x. A simple model may describe the overall reaction between TEOS and ozone as [25]:

$$Si(OC_2H_5)_4 + 4\,O_3 \rightarrow SiO_2 + 4\,CH_3CHO + 2\,H_2O + 4\,O_2$$

The coatings formed at different voltages are shown in Figure 5. Solid SiO_x are formed up to 10 kV, contrary to 12 kV when ozone and N_2O_5 are not present anymore downstream of the DBD (Table 2 and Figure 5, right). Similar homogeneous 2 nm coatings at 8 and 10 kV probably result from the similar ozone concentrations for both voltages (Figure 5, left and center).

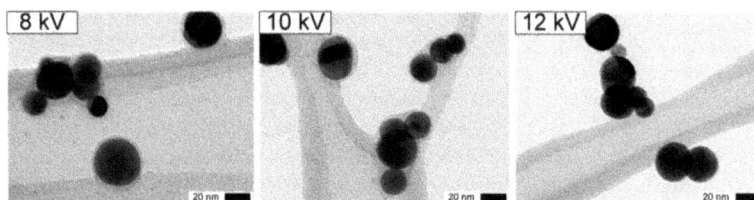

Figure 5. Transmission electron microscopy (TEM) micrograph of final solid coating of nanoparticles for different applied voltage (U_{pp}): (**left**) 8 kV; (**middle**) 10 kV; and (**right**): 12 kV, for 0.4 ppmv TEOS and $t_{Reaction}$ = 185 s.

The importance of ozone is further corroborated by a decrease in its concentration when TEOS is added to the system. FTIR measurements analogous to those described above yield an ozone concentration of only 55 ppmv at 10 kV, which is significantly lower than without precursor (301 ppmv). On the other hand, the N_2O_5 concentration of 19 ppmv is comparable to that measured without TEOS (24 ppmv). The addition of Pt particles to the system shows no further change in the ozone concentration (55 ppmv), but N_2O_5 decreases somewhat to 15 ppmv. While these findings suggest that ozone is the most important species for the reaction with TEOS in this process, further research is required to understand all possible reaction pathways.

Finally, since atomic oxygen can still be produced from slow post-DBD reactions of ozone and N_2O_5, it is not yet clear if TEOS reacts directly with ozone or with atomic oxygen to form condensable intermediate species.

3.2.1. Coating Composition

For two coating thicknesses controlled by the reaction time (Table 3), the particles were analyzed by EDX showing contributions of Pt (core particles) and Cu from the TEM grid and sample holder as well as C, O and Si. While Si and O obviously belonged to the SiO_x coatings, the source of C at least partly arises from the carbon coating of the TEM grid. As expected, the ratio of the coating elements to the core material increases for thicker coatings. The ratio of Si to O is nearly stoichiometric SiO_2 with a slight excess of oxygen. To determine the actual carbon content, coated Pt particles were deposited on a Mo TEM grid without a C film. The experiment showed that there is indeed some amount of carbon in the coated particles. The chemical composition of the coating following from this measurement is $SiO_{2.3}C_{1.5}H_z$. The H content cannot be measured with EDX.

3.2.2. Ranges of TEOS Concentration and Reaction Duration for Nanoparticle Coating

With the selected DBD arrangement polarized with applied voltages up to 10 kV at 3 L/min, different TEOS concentrations and reaction times have been tested. The goal was to define the ranges of conditions for the parametric study to control the thickness of nanometer sized coating, detailed in the next sections.

For longer reaction durations, a 20 L tank was used with a broad residence time distribution, varying from a few minutes up to several tens of minutes. With this tank downstream of the DBD and the highest TEOS concentrations (1400 ppmv), high concentrations of micron-sized spherical dense SiO_x particles were collected on TEM grids even without Pt nanoparticles from the spark generator (see Figure 6, left.). In that case, post-DBD heterogeneous nucleation of condensable gaseous intermediates probably occurs on single digit nano-sized particles and/or on ions reported downstream of DBDs [22,26,27]. Lowering the TEOS concentrations and reaction time lead to much smaller sub-micron-sized SiO_x coating (Figure 6, center), down to nanometer thick homogeneous coatings on 30 nm Pt nanoparticles (Figure 6, right).

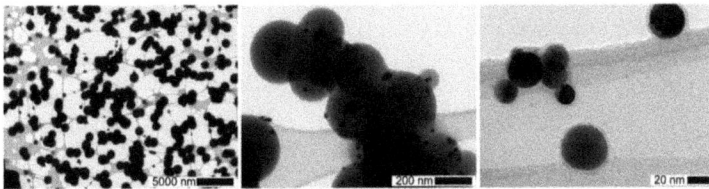

Figure 6. TEM micrographs: (**left**) Large SiO_x particles (1.13 µm) formed in the gas phase during the reaction duration in the 20 L tank (U_{pp} = 10 kV, with post-DBD injection of c_{TEOS} = 1400 ppmv without Pt nanoparticles); (**middle**) smaller sub-micron sized SiO_x coated Pt particles, formed and agglomerated in the gas phase (U_{pp} = 8.6 kV, 20 L tank with 280 ppmv TEOS and Pt nanoparticles); and (**right**) nanometer thick coatings of Pt nanoparticles (U_{pp} = 8 kV, $t_{Reaction}$ = 185 s with 0.4 ppmv TEOS).

For different reaction durations and TEOS concentrations, TEM and EDX analyses confirm that SiO_x particles are always formed in the gas phase, even if slow reactions of SiO_x formation still occur after collection near ambient temperature, as proved in the next section. Such spherical SiO_x particles usually arise from liquid droplets [4,7].

As the precursor concentration is below saturation (<0.2% of $p_{sat\ TEOS}$), these droplets cannot be formed by TEOS condensation. Indeed, the size distributions measured on line downstream of the reactor with SMPS, are the same for spherical sintered Pt nanoparticle with or without TEOS (see Figure 4, left). These droplets are thus formed by condensation of gaseous intermediate species continuously formed by TEOS reactions in post-DBD, up to the onset of condensation.

Hence, nanoparticles can be coated with SiO_x by a three step process starting with gas phase reactions of TEOS with post-DBD ozone and dinitrogen pentoxides, producing intermediate species more condensable than the TEOS precursor. In a second step, droplets are formed by condensation of silanol-like intermediates on nanoparticles, then converted by sol-gel reactions into solid SiO_x coatings. Finally, coagulation discussed below sometimes leads to agglomerated coated particles.

As expected from kinetic considerations, the growth rate of coating and the related thickness depend on the reaction conditions, including temperature and reactants concentrations downstream of the DBD (ozone and N_2O_5 plasma species, as well as TEOS). From a practical point of view, for fixed DBD operating conditions, the reaction time and the precursor concentration control the SiO_x coating thickness. To avoid the coating of the TEM grid, low TEOS concentrations in the ppmv range will be used. In this way, nanometer coating thickness can be achieved, as detailed in the next sections.

3.3. Homogeneous Nanometer-Sized Coating of Spherical Nanoparticles

To investigate the influence of particle properties and operational parameters on the coating thickness, size-selected sintered spherical Pt nanoparticles were used. TEOS concentrations and the reaction time were varied and the coating thickness was determined for different materials.

As outlined above, during the reaction of TEOS vapor with the plasma species intermediate molecules are formed which condense on the particle surfaces. Since the Knudsen number (*Kn*), defined as twice the mean free path ($\lambda \approx 65$ nm) of the gas molecules divided by the particle diameter, is significantly larger than 1 ($Kn \approx 4$), the condensation takes place close to the free molecular regime ($Kn \gg 1$). In this case the theory for heterogeneous condensation [28] predicts that the particle diameter increases linearly with the concentration of the condensing vapor and with the residence time as long as the concentration of the condensing vapor is constant, *i.e.*, without depletion:

$$\Delta R = \frac{M \cdot (p - p_S)}{\rho_l \cdot N_A \cdot \sqrt{2\pi mkT}} \cdot \Delta t \qquad (2)$$

where ΔR = coating thickness, M = molecular weight of the liquid, p = partial pressure of condensing vapor, p_S = partial pressure at particle surface, ρ_l = liquid density, N_A = Avogadro number, m = mass of vapor molecule, and kT = thermal energy.

3.3.1. Kinetics of TEOS Conversion into SiO_x Coating

Although the mechanisms and the kinetics of the formation reactions leading to the condensing species are not known in detail, our results support that such depletion of intermediate condensable species does not affect condensation here, even though it necessarily happens along the reaction volume, while post-DBD plasma species are consumed by reaction with TEOS. However, without a final forced condensation step by cooling, all formed intermediates do not condense. The results shown in Figure 4 and in Figure 8 confirm that free molecular condensation is a reasonable approximation to describe the thickness of the liquid coating *versus* TEOS concentration and reaction time, as detailed below.

The size distribution of particles/drops suspended in carrier gas was first measured with the SMPS at ambient conditions before collection, *i.e.*, for the reference time of reaction, while TEM analyses were performed on particles with residual solid coating only, formed after additional time delay of 2 min maximum for particles collected immediately and at least 20 min in air before final forced evaporation under vacuum conditions for TEM analysis. Therefore, the TEM results show the layers having a vapor pressure low enough to withstand the vacuum, *i.e.*, solid coatings.

Some examples of TEM micrographs are shown in Figure 7 for a TEOS concentration of 0.7 ppmv and different reaction times. While the solid coating for a reaction time of 83 s is only 2.5 nm thick, it increases to 38 nm after 185 s (Figure 8, right). For longer reaction times, agglomeration may take place as shown in Figure 7 (right). Agglomeration may occur by diffusion or by electrical forces between oppositely charged particles, since bipolar ions are produced downstream of such a DBD, which can even be used as an aerosol neutralizer [26,27].

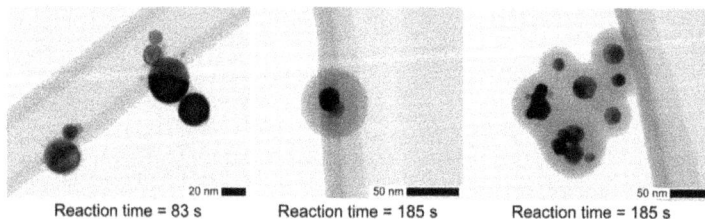

| 20 nm | 50 nm | 50 nm |
| Reaction time = 83 s | Reaction time = 185 s | Reaction time = 185 s |

Figure 7. TEM micrographs showing the time dependence of the coating thickness for a constant TEOS concentration of 0.7 ppmv.

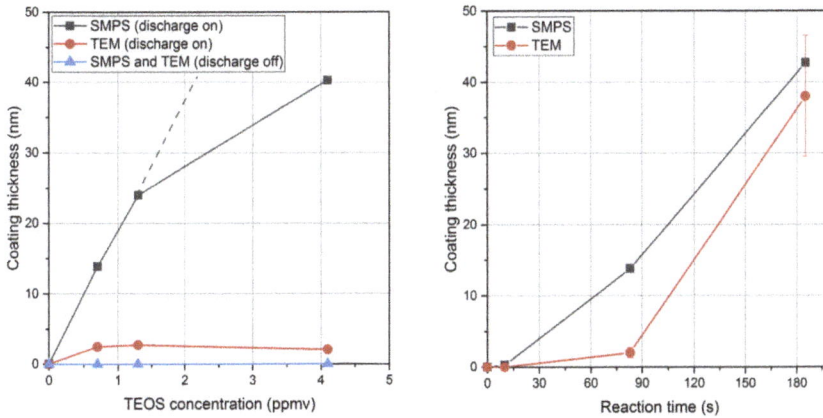

Figure 8. Coating thickness measured with scanning mobility particles sizer (SMPS) and TEM as a function of TEOS concentration for $t_{Reaction}$ = 83 s (**left**) and as a function of the reaction time for 0.7 ppmv TEOS concentration (**right**).

Particles and a few particle-clusters are well separated with smeared boundaries so that the coating thickness may only be deduced from the outside contour. This supports that core particles are not mobile in the coating, *i.e.*, the core is surrounded by a solid or at least more viscous gel shell. However, at the moment of collision the surface is liquid-like leading to conforming contacts. This is confirmed from the larger coating of drops/particles (measured before collection with SMPS) in comparison with the residual solid coating, formed after additional time delay before the final forced evaporation under vacuum conditions for TEM analysis (Figure 8), detailed below.

For a reaction time of 83 s, the coating thickness measured with SMPS increases with the TEOS concentration in qualitative agreement with Equation (1) (Figure 8, left). The deviation from the linear relationship at higher TEOS concentration may indicate that the concentration of the condensing vapor species is not directly proportional to the TEOS concentration. The reaction time of 83 s seems too short to complete the conversion of all the liquid coating into a solid silica coating. The reaction of formation of solid SiO_x is thus not limited by the amount of intermediate (unless much smaller precursor concentrations are used) but rather by a too short reaction time. Then, larger coatings could only be achieved for longer reaction times, as depicted below.

For longer reaction times, the thickness of the solid coating approaches the one of the liquid coating measured with SMPS (Figure 8, right). For a TEOS concentration of 0.7 ppmv the thickness of the liquid coating measured before collection increases with time corroborating Equation (1). An induction period is reported here before condensation linearly increases with time, as for the condensation of organic vapors on metal nanoparticles [29]. The solid coating also increases with time. However, reaction times of a few minutes are required to convert all the liquid intermediates into the final solid coating with the closest approach of thicknesses measured with SMPS and TEM, respectively. Conversion of liquid intermediates into solid SiO_x coatings might still occur after collection as already reported [25]. Actually, much slower rates of conversion are expected here with post-DBD reaction temperatures close to ambient temperature, than in classical sol-gel process for the formation of oxides usually performed above a few hundred degrees Celsius [4].

Hence, the first gas phase reaction and the second condensation step leading to suspended droplets of intermediates are faster than the final self-reaction of silanol-like intermediates leading to the solid silica coating, for the near ambient temperature of the reaction. The step limiting the overall reaction rate is the conversion of the condensed intermediates into solid silica coating. The thickness of final solid SiO_x coatings can be controlled from a nanometer to few tens of nanometers by playing on TEOS concentration for reaction times longer than 3 min, so that the SiO_x formation is nearly completed before collection, and on reaction times for TEOS concentrations in the ppmv range.

3.3.2. Influence of the Particle Material on the Coating Thickness

Different electrode materials (Pt, Cu, Ti, Ni, Fe) were used in the SDG producing different metal nanoparticles, sintered and coated without prior size classification. Contrary to the inert Pt, other metal nanoparticles were oxidized. A homogeneous SiO_x coating was created on spherical platinum, copper oxide and titanium oxide particles (Figure 9).

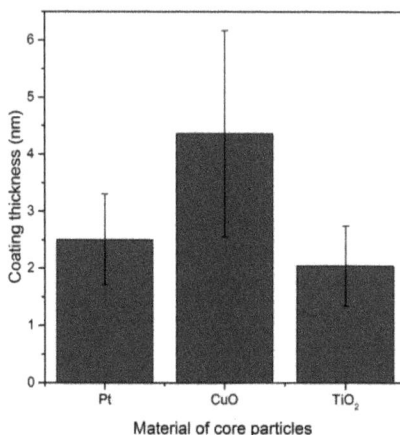

Figure 9. Influence of the particle material on the coating thickness from TEM analysis at a given TEOS concentration of 0.7 ppmv for $t_{Reaction}$ = 83 s.

For the nickel and iron oxide particles, complete coalescence to spherical particles was not achieved so that the coating was performed with agglomerates. A very thin SiO_x coating was found on the outer surface of these particles, while most of the SiO_x was deposited inside the agglomerates. Therefore, the thicknesses of such heterogeneous coatings on nickel and iron oxide agglomerates have not been plotted here. The coating of agglomerates will be discussed in the next section.

3.3.3. Influence of the Particle Concentration on the Coating Thickness

For given TEOS concentration assumed to be related to the gas phase concentration of intermediates (N_{vap}), the coating thickness could also be controlled *versus* the total concentration of particles ($N_{part.}$) affecting the amount of condensable species per particle ($N_{vap}/N_{part.}$). The depletion effect could only be addressed for longer reaction times (>3 min) and smaller TEOS concentrations with a final forced condensation. Similar coating thickness was found even though the particle number concentrations differed by about two orders of magnitude for classified and non-classified Pt particles and for the different materials tested.

3.4. Heterogenous Coatings of Agglomerates

While spherical nanoparticles facilitate the investigation of the coating kinetics ("idealized system"), nanoparticles in technical applications are mostly encountered as agglomerates consisting of much smaller primary particles as shown in Figure 2 (left). The concave surface areas in the necks between individual primary particles as well as in the interior of the highly ramified agglomerates are most prone for condensational effects inducing capillary condensation [29]. Therefore, the coating of nanoparticle agglomerates is a superposition of structure-induced deposition and reaction to form SiO$_x$ layers. Since deposition and coating inside the agglomerates will affect much less the agglomerate mobility compared to the coating of a sphere, the SMPS measurements can only provide a qualitative picture of the coating progress. Thus, another surface sensitive *in situ* technique is used here to follow the degree of surface coverage, *i.e.*, the aerosol photoemission (APE) [29]. Since SiO$_2$ has a high photothreshold (>10 eV [30]) that can only be overcome by extreme short wavelength UV light (far below UV-C), silica coatings are commonly employed to inhibit photocatalytic effects of sun blocking nanoparticles such as ZnO or TiO$_2$ [1] by suppressing the exchange of photogenerated species with the environment [2].

As starting point for the coating of Pt nanoparticle agglomerates the condensation of pure TEOS vapor without plasma (Figure 10, left) and with plasma turned on (Figure 10, right) was investigated as a function of the precursor concentration by tandem DMA measurements.

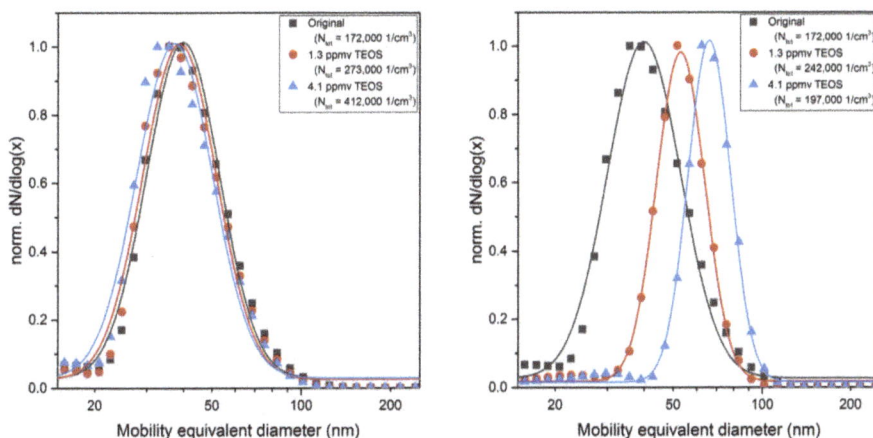

Figure 10. Change in mobility equivalent diameter depending on the precursor concentration without (**left**) and with (**right**) plasma for agglomerates t_{Reaction} = 83 s).

As shown in Figure 10 (left), small variations of the agglomerate size distribution were observed without a pronounced shift when the plasma was turned off. However, while with the plasma turned on a systematic shift of the size similar to the behavior of spherical particles was observed, the equivalent layer thickness appears to be smaller in comparison. On the one hand, the available surface area for the agglomerates is substantially higher (about a factor of 3) compared to the spherical Pt particles due to surface losses of the Pt particles as a consequence of coalescence but also due to particle losses in the sintering furnace. Therefore, the amount of condensing species per particle surface favors thicker layers on spherical particles. On the other hand, as outlined above, substantial condensation occurs inside agglomerates with a minor effect on the measured agglomerate mobility.

Next, the morphology of the solid silica coatings of the Pt agglomerates was studied with TEM for different TEOS concentrations and different reaction times. The coating in Figure 11 covers nearly the whole agglomerate. As already found for the spherical particles, the coating thickness of the agglomerates depends more on the reaction time than on the precursor concentration (*cf.* Figure 8).

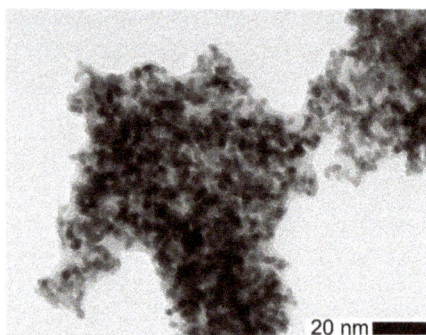

Figure 11. TEM micrograph of a coated Pt agglomerate after a reaction time of 185 s with a TEOS concentration of 0.7 ppmv showing a thick coating.

In order to follow the coating kinetics, photoemission measurements were done for Pt agglomerates as well as spheres as a function of the precursor concentration (Figure 12). The non-coated agglomerates ($c_{TEOS} = 0$) gave the highest signal corresponding to the largest amount of active surface area. Uncoated sintered spheres gave a smaller signal due to a reduced particle surface area, a consequence of particle losses (number concentration) and surface loss due to sintering. For the spheres, a small amount of TEOS is sufficient to obtain a complete coating, reflected in a rapid drop of the APE signal. Higher precursor concentrations increased the coating thickness, but even a small, hermetic coating was enough to prevent photoemission. Agglomerates, however, behaved differently in the way that coating was favored in the necks and the gaps between branches, leaving some outer particles exposed. The fraction of the uncoated primary particles contributing to the APE signal decreased with increasing TEOS concentration. At high TEOS concentrations the whole agglomerate was covered. Parallel to the complete coating of the agglomerates a restructuring of some branches to more compact structures was observed due to capillary forces. Such a compactization of nanoparticle agglomerates is well documented in the literature [31]. However, at lower precursor concentrations, restructuring did not occur. This observation raises the question if the mere coating of the necks would lead to mechanical stabilization against restructuring without hindering mass transport processes, which would enable interesting applications such as catalysis at higher temperatures.

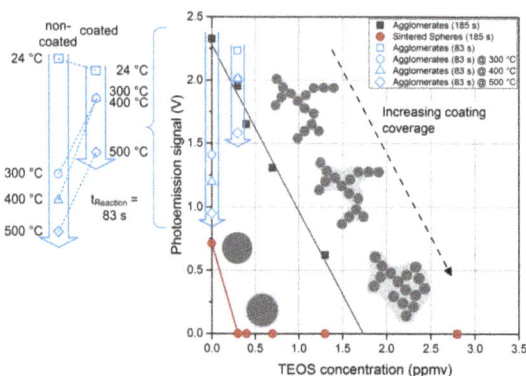

Figure 12. Measurement of the coating coverage by photoemission for Pt particles and sketches of the different kinds of observed coatings from TEM micrographs ($t_{Reaction} = 185$ s). Additionally, the open, blue symbols show the change in APE signal depending on the sintering temperature, analogous to the particles shown in Figure 13 ($t_{Reaction} = 83$ s).

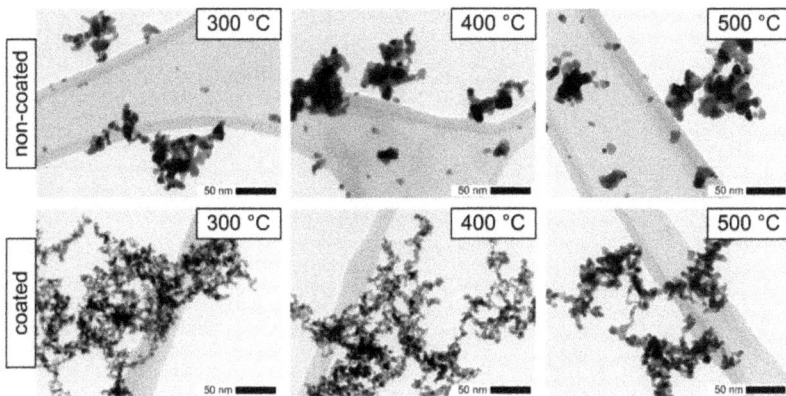

Figure 13. TEM micrographs showing the sintering behavior of non-coated (**top**) and coated (**bottom**) Pt particles during the transit through a tube furnace (residence time at 24 °C of 8 s).

Silica coatings have been employed in various forms to protect noble metal nanoparticle catalysts from sintering [32]. In this context, two opposing effects have to be managed: On the one hand, the thermal stability of the nanoparticles increases with increasing coating thickness. On the other hand, thicker silica layers hamper the mass transport across the coating reducing the catalytic activity. To overcome this problem, one solution is to obtain a sufficient porosity of the coating by high temperature treatment [1] or by adding pore building agents [32]. However, an easier and more versatile approach would be the precise control of the coating process by depositing silica only in the necks between the primary particles. Such a technique was applied by Kim and Ehrman [33] using TEOS as precursor to coat TiO_2 nanoparticles.

To test if such a precise control of the coating can also be realized in the post-plasma process investigated here minute amounts of TEOS precursor (0.3 ppmv) and short reaction times (83 s) were used. As a measure for the surface coverage, the APE signal of the coated Pt agglomerates was determined and compared to the uncoated agglomerates. As indicated in Figure 12 (open, blue symbols), the photoemission activity was only reduced by about 5% due to the silica coating (silica confirmed by EDX). It was verified by TEM microscopy that at these conditions most of the silica was in fact deposited in the necks between the primary particles leaving most of the Pt surface accessible for interaction with the surrounding gas molecules. In order to test if this coating was sufficient to stabilize the agglomerates against thermal restructuring, the sintering behavior of non-coated and coated particles was studied by passing the aerosol particles through a tube furnace at a constant gas flow rate either directly after the SDG or after coating. Figure 13 shows the TEM micrographs for furnace temperatures of 300, 400 and 500 °C, respectively. The applied coating, which conserved the surface functionality as shown above, was very thin and non-homogeneous. With increasing temperature, the differences in the morphologies of the coated and non-coated agglomerates became more pronounced. While the primary particles of the non-coated agglomerates visibly sintered to larger diameters even at temperatures as low as 300 °C, the coated particles showed much smaller changes and retained a branched structure. The thin coating did not completely prevent the sintering of primary particles but it preserved a higher amount of surface area. As indicated by the blue symbols in Figure 12, also the photoactivity of the coated Pt agglomerates was only reduced by 12% for temperatures up to 400 °C and by 29% for a temperature of 500 °C. The uncoated agglomerates start with a slightly higher initial photoactivity at room temperature (24 °C), which decreases at elevated temperatures much faster compared to the coated particles (*cf.* insert in Figure 12). At 500 °C the remaining photoactivity of the coated particles is about 60% higher than the one of the uncoated Pt agglomerates. However, the coating process was not optimized so far and additional experiments will be necessary to explore the potential and to define the limits of this coating process for catalyst particles. Nevertheless, the first results of the post-plasma coating are promising and suggest further applications. Indeed, the operating conditions can be tuned to cover preferentially the interparticle contact zones between primary particles within agglomerates. This hampers the sintering of platinum agglomerates, retaining high apparent surface to much higher temperatures than usual sintering temperatures (e.g., for TiO_2 agglomerates that could be heated more than 400 °C above the usual sintering temperature [1]).

4. Conclusions

Homogeneous SiO_x coatings of spherical and agglomerated metal and metal oxide nanoparticles (Pt, CuO, and TiO_2) have been achieved in post-DBD. EDX results suggest some amount of carbon in the coating.

It has been shown that non-thermal plasma filaments induced in a dielectric barrier discharge (DBD) trigger post-DBD gas phase reactions of ozone and dinitrogen pentoxide with gaseous TEOS precursor, forming the final homogeneous SiO_x coatings.

This confirms the interest of post-DBD injection of the organo-silicon precursor to avoid electrode coating to achieve stable DBD production of active species with subsequent controlled thickness of SiO_x coatings of nanoparticles *versus* TEOS concentration and reaction time.

The coating thickness can be controlled through the reaction time and the precursor concentration. At this point, it has to be underlined that larger coating thicknesses were measured before collection in the gas phase from size distribution measurements than after collection by TEM analysis. A possible explanation is the formation of a liquid intermediate product, while the subsequent reaction to a solid SiO_x coating requires more time and could still happen after collection.

The process seems independent of the particle material, so that very homogeneous coatings can be achieved on nanoparticles from different materials and with different structures (single spherical particles as well as agglomerates). Even thermally unstable materials could be coated at near ambient temperatures.

For agglomerates, the operating conditions can be tuned to cover preferentially the interparticle contact zones between primary particles. This hampers the sintering of platinum agglomerates, retaining high apparent surface to much higher temperatures than usual sintering temperatures. The coatings are suitable for enhanced thermal stability, as well as for tunable photoactivity.

Post-plasma chemistry is complex because of numerous possible pathways, and more analysis in different plasma and mixing conditions would be required to identify all reaction pathways. Then, the growth rate of the coating and the related coating thickness could be defined *versus* conditions of reaction (temperature and concentrations of reactants—plasma species and TEOS—as well as the relative concentration of intermediate condensable species and nanoparticles—$N_{condensing\ vap}/N_{part}$), probably also controlling the amount of liquid condensed per particle, and thus the final solid coating thickness.

Acknowledgments: The financial support by the German Research Foundation under grant WE 2331/18-1 is gratefully acknowledged.

Author Contributions: P.P. performed the experiments. N.J. and J.-P.B. characterized the plasma. P.P., J.-P.B. and A.P.W. conceived and designed the experiments, analyzed the data and wrote the paper.

Conflicts of Interest: The authors declare no conflict of interest.

References

1. Qi, F.; Moiseev, A.; Deubener, J.; Weber, A. Thermostable photocatalytically active TiO_2 anatase nanoparticles. *J. Nanopart. Res.* **2011**, *13*, 1325–1334. [CrossRef]

2. Jansen, R.; Osterwalder, U.; Wang, S.Q.; Burnett, M.; Lim, H.W. Photoprotection: Part II. Sunscreen: Development, efficacy, and controversies. *J. Am. Acad. Dermatol.* **2013**, *69*. [CrossRef] [PubMed]

3. Graf, C.; Vossen, D.L.J.; Imhof, A.; van Blaaderen, A. A General Method to Coat Colloidal Particles with Silica. *Langmuir* **2003**, *19*, 6693–6700. [CrossRef]

4. Kodas, T.T.; Hampden-Smith, M.J. *Aerosol Processing of Materials*; Wiley-VCH: New York, NY, USA, 1999.

5. George, S.M. Atomic Layer Deposition: An Overview. *Chem. Rev.* **2010**, *110*, 111–131. [CrossRef] [PubMed]

6. Teleki, A.; Heine, M.C.; Krumeich, F.; Akhtar, M.K.; Pratsinis, S.E. *In Situ* Coating of Flame-Made TiO$_2$ Particles with Nanothin SiO$_2$ Films. *Langmuir* **2008**, *24*, 12553–12558. [CrossRef] [PubMed]

7. Park, H.K.; Park, K.Y. Control of Particle Morphology and Size in Vapor-Phase Synthesis of Titania, Silica and Alumina Nanoparticles. *KONA Powder Part. J.* **2015**, *32*, 85–101. [CrossRef]

8. Massines, F.; Sarra-Bournet, C.; Fanelli, F.; Naudé, N.; Gherardi, N. Atmospheric Pressure Low Temperature Direct Plasma Technology: Status and Challenges for Thin Film Deposition. *Plasma Process. Polym.* **2012**, *9*, 1041–1073. [CrossRef]

9. King, D.M.; Liang, X.; Burton, B.B.; Kamal Akhtar, M.; Weimer, A.W. Passivation of pigment-grade TiO$_2$ particles by nanothick atomic layer deposited SiO$_2$ films. *Nanotechnology* **2008**, *19*. [CrossRef] [PubMed]

10. Boies, A.M.; Calder, S.; Agarwal, P.; Lei, P.; Girshick, S.L. Chemical Kinetics of Photoinduced Chemical Vapor Deposition: Silica Coating of Gas-Phase Nanoparticles. *J. Phys. Chem. C* **2012**, *116*, 104–114. [CrossRef]

11. Jiménez, C.; De Barros, D.; Darraz, A.; Deschanvres, J.-L.; Rapenne, L.; Chaudouët, P.; Méndez, J.E.; Weiss, F.; Thomachot, M.; Sindzingre, T.; *et al.* Deposition of TiO$_2$ thin films by atmospheric plasma post-discharge assisted injection MOCVD. *Surf. Coat. Technol.* **2007**, *201*, 8971–8975. [CrossRef]

12. Borra, J.-P.; Valt, A.; Arefi-Khonsari, F.; Tatoulian, M. Atmospheric Pressure Deposition of Thin Functional Coatings: Polymer Surface Patterning by DBD and Post-Discharge Polymerization of Liquid Vinyl Monomer from Surface Radicals. *Plasma Process. Polym.* **2012**, *9*, 1104–1115. [CrossRef]

13. Fanelli, F.; Fracassi, F. Aerosol-Assisted Atmospheric Pressure Cold Plasma Deposition of Organic–Inorganic Nanocomposite Coatings. *Plasma Chem. Plasma Process.* **2014**, *34*, 473–487. [CrossRef]

14. Mori, T.; Tanaka, K.; Inomata, T.; Takeda, A.; Kogoma, M. Development of silica coating methods for powdered pigments with atmospheric pressure glow plasma. *Thin Solid Films* **1998**, *316*, 89–92. [CrossRef]

15. Kogoma, M.; Tanaka, K.; Takeda, A. Powder Treatments using Atmospheric Pressure Glow Plasma (Silica Coating of TiO$_2$ Fine Powder). *J. Photopolym. Sci. Technol.* **2005**, *18*, 277–280. [CrossRef]

16. Brueser, V.; Hahnel, M.; Kersten, H. Thin Film Deposition on Powder Surfaces Using Atmospheric Pressure Discharge. In *New Vistas in Dusty Plasmas: Fourth International Conference on the Physics of Dusty PlasmasAIP Conference Proceedings*; American Institute of Physics: College Park, MD, USA, 2005; Volume 799, pp. 343–346.

17. Vons, V.; Creyghton, Y.; Schmidt-Ott, A. Nanoparticle production using atmospheric pressure cold plasma. *J. Nanopart. Res.* **2006**, *8*, 721–728. [CrossRef]

18. Nessim, C.; Boulos, M.; Kogelschatz, U. In-flight coating of nanoparticles in atmospheric-pressure DBD torch plasmas. *Eur. Phys. J. Appl. Phys.* **2009**, *47*. [CrossRef]

19. Meuller, B.O.; Messing, M.E.; Engberg, D.L.J.; Jansson, A.M.; Johansson, L.I.M.; Norlén, S.M.; Tureson, N.; Deppert, K. Review of Spark Discharge Generators for Production of Nanoparticle Aerosols. *Aerosol Sci. Technol.* **2012**, *46*, 1256–1270. [CrossRef]

20. Borra, J.-P.; Jidenko, N.; Hou, J.; Weber, A. Vaporization of bulk metals into single-digit nanoparticles by non-thermal plasma filaments in atmospheric pressure dielectric barrier discharges. *J. Aerosol Sci.* **2015**, *79*, 109–125. [CrossRef]

21. Weber, A.P.; Seipenbusch, M.; Kasper, G. Application of Aerosol Techniques to Study the Catalytic Formation of Methane on Gasborne Nickel Nanoparticles. *J. Phys. Chem. A* **2001**, *105*, 8958–8963. [CrossRef]

22. Borra, J.-P. Nucleation and aerosol processing in atmospheric pressure electrical discharges: Powders production, coatings and filtration. *J. Phys. Appl. Phys.* **2006**, *39*, R19–R54. [CrossRef]

23. Okuyama, K.; Fujimoto, T.; Hayashi, T.; Adachi, M. Gas-phase nucleation in the tetraethylorthosilicate (TEOS)/O$_3$ APCVD process. *AIChE J.* **1997**, *43*, 2688–2697. [CrossRef]

24. Fujino, K.; Nishimoto, Y.; Tokumasu, N.; Maeda, K. Silicon Dioxide Deposition by Atmospheric Pressure and Low-Temperature CVD Using TEOS and Ozone. *J. Electrochem. Soc.* **1990**, *137*, 2883–2887. [CrossRef]

25. Romet, S.; Couturier, M.F.; Whidden, T.K. Modeling of silicon dioxide chemical vapor deposition from tetraethoxysilane and ozone. *J. Electrochem. Soc.* **2001**, *148*, G82–G90. [CrossRef]

26. Bourgeois, E.; Jidenko, N.; Alonso, M.; Borra, J.P. DBD as a post-discharge bipolar ions source and selective ion-induced nucleation *versus* ions polarity. *J. Phys. Appl. Phys.* **2009**, *42*. [CrossRef]
27. Mathon, R.; Jidenko, N.; Borra, J.-P. Characterization of a Post-DBD Aerosol Bipolar Diffusion Neutralizer for SMPS Size Distribution Measurements. In Proceedings of 2016 European Aerosol Conference, Tours, France, 4–6 September 2016.
28. Hinds, W.C. *Aerosol Technology: Properties, Behavior, and Measurement of Airborne Particles*, 2nd ed.; Wiley: New York, NY, USA, 1999.
29. Poostforooshan, J.; Rennecke, S.; Gensch, M.; Beuermann, S.; Brunotte, G.-P.; Ziegmann, G.; Weber, A.P. Aerosol Process for the *In Situ* Coating of Nanoparticles with a Polymer Shell. *Aerosol Sci. Technol.* **2014**, *48*, 1111–1122. [CrossRef]
30. Wellert, S. Wechselwirkung von Elektronen und Molekülen Mit Einzelnen SiO_2-Nanopartikeln: Massenanalyse in Einer Vierpolfalle. Ph.D. Thesis, Fakultät für Naturwissenschaften, Technische Universität Chemnitz, Chemnitz, Germany, 2003.
31. Miljevic, B.; Surawski, N.C.; Bostrom, T.; Ristovski, Z.D. Restructuring of carbonaceous particles upon exposure to organic and water vapours. *J. Aerosol Sci.* **2012**, *47*, 48–57. [CrossRef]
32. Park, J.-N.; Forman, A.J.; Tang, W.; Cheng, J.; Hu, Y.-S.; Lin, H.; McFarland, E.W. Highly Active and Sinter-Resistant Pd-Nanoparticle Catalysts Encapsulated in Silica. *Small* **2008**, *4*, 1694–1697. [CrossRef] [PubMed]
33. Kim, S.; Ehrman, S.H. Capillary Condensation onto Titania (TiO_2) Nanoparticle Agglomerates. *Langmuir* **2007**, *23*, 2497–2504. [CrossRef] [PubMed]

nanomaterials

MDPI

Communication

Highly-Efficient Plasmon-Enhanced Dye-Sensitized Solar Cells Created by Means of Dry Plasma Reduction

Van-Duong Dao and Ho-Suk Choi *

Department of Chemical Engineering, Chungnam National University, 220 Gung-Dong, Yuseong-Gu,
Daejeon 305-764, Korea; duongdaovan@cnu.ac.kr or duongdaovan@gmail.com
* Correspondence: hchoi@cnu.ac.kr; Tel.: +82-42-821-5689

Academic Editors: Krasimir Vasilev and Melanie Ramiasa
Received: 2 February 2016; Accepted: 9 April 2016; Published: 14 April 2016

Abstract: Plasmon-assisted energy conversion is investigated in a comparative study of dye-sensitized solar cells (DSCs) equipped with photo-anodes, which are fabricated by forming gold (Au) and silver (Ag) nanoparticles (NPs) on an fluorine-doped tin oxide (FTO) glass surface by means of dry plasma reduction (DPR) and coating TiO_2 paste onto the modified FTO glass through a screen printing method. As a result, the FTO/Ag-NPs/TiO_2 photo-anode showed an enhancement of its photocurrent, whereas the FTO/Au-NPs/TiO_2 photo-anode showed less photocurrent than even a standard photo-anode fabricated by simply coating TiO_2 paste onto the modified FTO glass through screen printing. This result stems from the small size and high areal number density of Au-NPs on FTO glass, which prevent the incident light from reaching the TiO_2 layer.

Keywords: plasmonic; dry plasma reduction; Ag nanoparticles; Au nanoparticles; dye-sensitized solar cells

1. Introduction

Dye-sensitized solar cells (DSCs) can achieve comparatively high conversion efficiency levels at a low fabrication cost [1]. In general, a DSC has a working electrode (WE) in the form of a nano-crystalline TiO_2 electrode modified with a dye and fabricated on a transparent conducting oxide (TCO), an iodide electrolyte solution and a platinum (Pt) counter electrode (CE). The WE has the greatest potential to improve the efficiency of a DSC, with the efficiency defined as the light-harvesting efficiency, due to its monolayer of adsorbed dye molecules [2]. In order to improve the photocurrent, tremendous efforts have been made, including efforts to control the thickness of the TiO_2 layer [3], synthesize new dyes [4] or use a scattering layer [5]. A major problem, however, is the attachment of less dye, reducing the short-circuit current density [6]. One approach to solve this problem is based on a plasmonic structure, such as silver nanoparticles (Ag-NPs) [6,7], gold nanoparticles (Au-NPs) [8] or gold nanorods [9]. Thus far, plasmonic structures have been prepared by sputtering combined with annealing [6], soaking in a colloidal silver solution for as many as 15 hours [7] or with a plasmonic paste [8,9]. These techniques, however, require toxic chemicals for the synthesis process, high temperatures for annealing, extended times for soaking or a vacuum chamber for sputtering, all indicating that developing an economic continuous process remains a challenge [10].

In order to overcome these process restrictions, we developed a new process of efficiently synthesizing supported metal-NPs by means of dry plasma reduction (DPR) [10–12]. More importantly, the developed method can operate at temperatures close to room temperature under atmospheric pressure and without using toxic chemicals. In this communication, we use the DPR method to synthesize Au-NPs and Ag-NPs on fluorine-doped tin oxide (FTO) glass substrates. The main goal of this study is to improve photocurrent generation in DSCs by studying the plasmon enhancement effect. This work is not only for developing a simple strategy of realizing how to improve photovoltaic properties of the DSC using NPs-coated FTO, but also for testing a potential concept of other applications, such as the optical devices, chemical catalysts and biological sensors.

2. Results

The overall DPR fabrication process is thoroughly described in our previous study [10]. Figure 1a,b shows the morphologies of the Ag-NPs and Au-NPs on the FTO glass substrates. As can be seen, Ag-NPs and Au-NPs with a few nanometers in size can clearly be observed to be uniformly formed on the FTO glass surfaces without any aggregation via the DPR process. This uniform and dense distribution of NPs provides high optical transmittance in the NPs on the FTO glass. As shown in Figure 1a,b, the Ag-NPs appear larger than the Au-NPs, while the areal number density of the Ag-NPs is lower than that of Au-NPs on the FTO glass. This is also supported by the Transmission Electron Microscopy (TEM) images of the Ag-NPs/Cu grid (Figure S1a, Supplementary Materials) and Au-NPs/Cu grid samples (Figure S1b, Supplementary Materials). Figure S1a shows a TEM image of Ag-NPs immobilized on a Cu grid. The average size of the Ag-NPs was measured and found to be 6–7 nm. Figure S1b presents a TEM image of Au-NPs immobilized on a Cu grid. The average size of the Au-NPs was in the range of 2–7 nm, typically 3 nm and quite small and highly mono-dispersed. Note that the particle sizes have been estimated through the ImageJ program. The average sizes of the Ag-NPs and Au-NPs on the Cu grid are slightly larger than that of the Pt-NPs on the Cu grid [10]. The size difference between the NPs corresponding to the different metal precursors on the Cu grids is likely due to the different electrostatic energy levels of the metal NPs. The physiochemical conversions of $HAuCl_4$ and $AgNO_3$ into metallic Au and Ag, respectively, were further confirmed by X-ray photoelectron spectroscopy (XPS) (Supplementary Materials, Figures S2 and S3). As shown in Figure S2, the binding energy levels of the Au $4f_{7/2}$ and Au $4f_{5/2}$ electrons are 83.9 and 87.7 eV, respectively. This is consistent with Au metal, which indicates that the Au-NPs/FTO sample exists in a metallic state [13]. Additionally, after DPR, the Cl element composition was completely absent in the XPS spectrum (Supplementary Materials, Table S1). Hence, $HAuCl_4$ was completely converted into metallic Au through the DPR process. It was also confirmed that Au-NPs were completely crystallized. The binding energy levels of the Ag $3d_{5/2}$ and Ag $3d_{3/2}$ electrons are 368.2 and 374.2 eV, respectively (Supplementary Materials, Figure S3), consistent with Ag metal and indicating that the Ag-NPs/FTO samples are in a metallic state [14]. Identical to the Cl element composition in H_2PtCl_6 [10] and $HAuCl_4$, the N element composition in $AgNO_3$ was also completely absent after DPR (Supplementary Materials, Table S2).

(a)   (b)

Figure 1. (a) High-Resolution Scanning Electron Microscopy (HRSEM) image of Ag-nanoparticles (NPs) on an fluorine-doped tin oxide (FTO) glass substrate; and (b) HRSEM image of Au-NPs on an FTO glass substrate.

Figure 2 shows a comparison of the performance capabilities of the three DSCs with different working electrodes (WEs). The DSC with FTO/Ag-NPs/TiO$_2$ as the WE shows an energy conversion efficiency of 7.49 ($\pm$0.14)%, which is better than those of the DSCs with the FTO/TiO$_2$ and FTO/Ag-NPs/TiO$_2$ WEs, videlicet 6.54 ($\pm$0.15)% and 6.27 ($\pm$0.11)%, respectively, as listed in Table 1. Regardless of the illumination condition, the open-circuit voltages, V_{oc}, of the DSCs with FTO/Ag-NPs/TiO$_2$ and FTO/Au-NPs/TiO$_2$ as the WEs are slightly higher than that of the DSC with FTO/TiO$_2$ as the WE. This phenomenon is understood by considering that the Ag-NPs and Au-NPs are linked together with titania, causing a shift of the conduction band edge of titania to the negative side [15] and further increasing the gap between the Fermi level of the photoelectrode and the redox potential under illumination, which is the definition of photovoltage [16]. The short-circuit current, J_{sc}, of the DSC with the FTO/Ag-NPs/TiO$_2$ WE, however, is slightly higher than that of the DSC with FTO/TiO$_2$ as the WE due to the low electron transfer at the TiO$_2$/dye/electrolyte interface [9], which was confirmed again by incident photon-to-current efficiency (IPCE) measurements and the electrochemical impedance spectroscopy (EIS) results below. In contrast, the J_{sc} of the DSC with FTO/TiO$_2$ as the WE is slightly higher than that of the DSC with FTO/Au-NPs/TiO$_2$ as the WE due to the higher Schottky barrier in the interfacial region of Au (5.1–5.5 eV) and TiO$_2$ (4.0 eV), with a slight difference between the work function of Ag (4.12 eV) and the electron affinity of TiO$_2$ (4.0 eV). There were some conflicts between our results and those by Lin *et al.* [6] for FTO/Ag-NPs/TiO$_2$ and by Zhang *et al.* [8] for FTO/Au-NPs/TiO$_2$. The differences in the NP size and the dense distribution were used to explain these phenomena. Indeed, the larger size and highly dense distribution of Ag-NPs mean that they could not preferentially attach to specific sites on the TiO$_2$ surface, resulting in a decrease of the catalytic activity. In contrast, the small size and highly dense distribution of Au-NPs on FTO glass would prevent the incident light from reaching the TiO$_2$ layer. The fill factor (FF) of the DSC with the FTO/Ag-NPs/TiO$_2$ WE (67.60% $\pm$ 0.68%) is higher than those of the DSCs with the FTO/TiO$_2$ WE (61.71% $\pm$ 1.16%) and the FTO/Au-NPs/TiO$_2$ WE (60.60% $\pm$ 1.02%) due to the low internal resistance of the cell [17]. This becomes clear when examining EIS results, as shown in Figure 3.

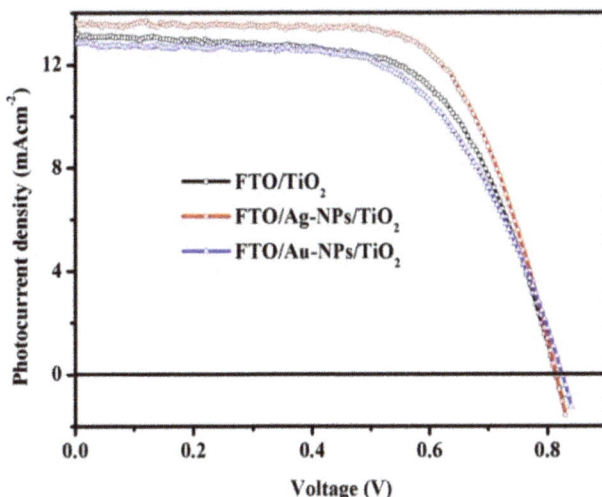

Figure 2. Current-voltage characteristics of three dye-sensitized solar cells (DSCs) equipped with different working electrodes.

Table 1. Photoelectric performance of the three cells shown in Figure 2. J_{sc}: The short-circuit current; V_{oc}: the open-circuit voltages; FF: Fill Factor; η: efficiency; FTO: fluorine-doped tin oxide; NPs: nanoparticles.

Working Electrode	J_{sc} $(mA \cdot cm^{-2})$	V_{oc} (mV)	FF (%)	η (%)
FTO/TiO$_2$	12.83 ± 0.21	820.00 ± 5.00	61.71 ± 1.16	6.54 ± 0.15
FTO/Ag-NPs/TiO$_2$	13.49 ± 0.16	821.66 ± 11.5	67.60 ± 0.68	7.49 ± 0.14
FTO/Au-NPs/TiO$_2$	12.34 ± 0.67	826.67 ± 7.63	60.60 ± 1.02	6.27 ± 0.11

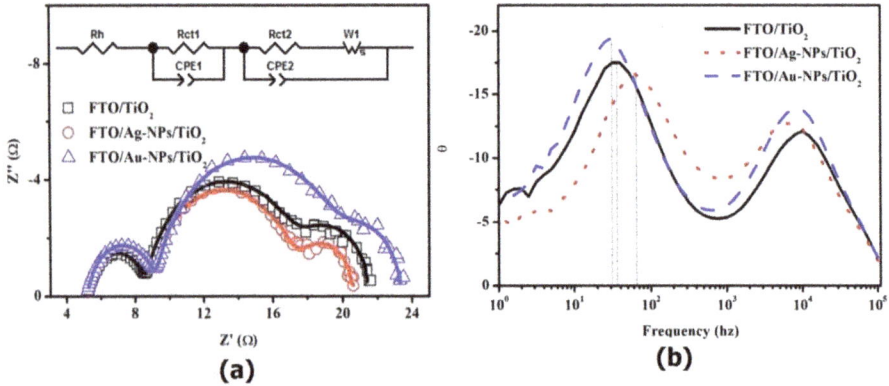

Figure 3. (**a**) Nyquist plots of three DSCs equipped with different working electrodes; and (**b**) The electrochemical impedance spectroscopy (EIS) Bode plots of three DSCs equipped with different working electrodes. Z": imaginary part of impedance; Z': real part of impedance; CPE1: the constant phase element at counter electrode/electrolyte interface; CPE2: the constant phase element at working electrode/electrolyte interface; Rh: Ohmic internal resistance; Rct1: charge-transfer resistance at counter electrode/electrolyte interface; Rct2: charge-transfer resistance at working electrode/electrolyte interface; W1: Warburg impedance.

Figure 3a presents three Nyquist plots of DSCs with FTO/TiO$_2$, FTO/Ag-NPs/TiO$_2$ and FTO/Au-NPs/TiO$_2$ WEs. Table 2 shows the estimated charge transfer resistance values on both the CE and WE, as well as the constant phase element (CPE) parameters for the interfaces of both the CE and WEs. For the purpose of comparing only the WEs, we fabricated three CEs using the same materials and method. As shown in Table 2, none of the parameters pertaining to the CE interface showed a significant difference with regard to the three DSCs. However, there was a distinct difference in Rct2 among DSCs with different WEs, as shown in Table 2. The Rct2 value of the FTO/Ag-NPs/TiO$_2$ sample was low, at 3.93 Ω·cm^2, while those of the FTO/TiO$_2$ and FTO/Au-NPs/TiO$_2$ samples were 4.24 Ω·cm^2 and 5.23 Ω·cm^2, respectively. For our Au-NPs immobilized on FTO, there is direct contact between the Au and the electrolyte, with the surface of the Au-NPs serving as a recombination site for the photogenerated electrons and triiodide ions, resulting in an increase in the charge transfer resistance in comparison to the FTO/TiO$_2$ WE [8]. It is known that a lower Rct2 of a cell means faster redox kinetics of I_3^-/I^- pairs at the photoanode/dye/electrolyte interface [18]. Hence, the electron transfer from the electrolyte to the oxidized dye became rapid, suggesting an efficient injection of electrons to TiO$_2$ and transfer across the film to the outer circuit, reducing the rate of electron recombination [18]. This positive factor results in a high IPCE and higher power conversion efficiency. Furthermore, the reduction of the total internal resistance due to the decrease in Rct2 is the cause of the increase in the fill factor of the DSCs [17], which confirms the increase in the conversion efficiency, as previously shown.

Table 2. Impedance parameters of three DSCs with FTO/TiO$_2$, FTO/Ag-NPs/TiO$_2$ and FTO/Au-NPs/TiO$_2$ working electrodes (WEs), as estimated from the impedance spectra and equivalent circuit shown in Figure 3a. Rh: Ohmic internal resistance; Rct1: charge-transfer resistance at counter electrode/electrolyte interface; Rct2: charge-transfer resistance at working electrode/electrolyte interface; W1: Warburg impedance; CPE1: the constant phase element at counter electrode/electrolyte interface; CPE2: the constant phase element at working electrode/electrolyte interface. Note that CPE = (CPE-T)$^{-1}(jw)^{-(CPE-P)}$, in which $j^2 = -1$, w = frequency, CPE-T and CPE-P are frequency-independent parameters of the CPE. Ws = $R \times$ tanh([$j \times T \times w]^P)/(j \times T \times w)^P$, where R is real part of impedance; T is the diffusion interpretation of Warburg element and it is given by $T = L^2/D$ in which L is effective diffusion thickness and D is diffusion coefficient of particles; $P = 0.5$.

Working Electrode	Rh ($\Omega \cdot$ cm^2)	Rct1 ($\Omega \cdot$ cm^2)	CPE1-T (μF$\cdot$ cm^{-2})	CPE1-P	Rct2 ($\Omega \cdot$ cm^2)	Ws			CPE2-T (μF$\cdot$ cm^{-2})	CPE2-P
						R	T	P		
FTO/TiO$_2$	2.59	1.81	35.5	0.90	4.24	2.08	0.42	0.5	2410	0.90
FTO/Ag-NPs/TiO$_2$	2.58	1.85	36.9	0.89	3.93	1.73	0.43	0.5	2470	0.90
FTO/Au-NPs/TiO$_2$	2.56	1.89	36.1	0.90	5.23	1.74	0.40	0.5	2450	0.90

For more information about the electron recombination lifetime of the photo-injected electrons during the photovoltaic process ($\tau_e = 1/2\pi f_{max}$; f_{max} is the maximum frequency of the peaks in the intermediate frequency region of EIS [19,20]), EIS Bode plots were devised. We found that the highest and lowest f_{max} values were obtained with DSCs based on FTO/Ag-NPs/TiO$_2$ and FTO/Au-NPs/TiO$_2$ WEs, respectively. Therefore, the shortest and longest τ_e distances were obtained for DSCs based on FTO/Ag-NPs/TiO$_2$ and FTO/Au-NPs/TiO$_2$ WEs, respectively. It is well known that a longer τ_e indicates greater effective inhibition of electron recombination during the electron transfer process across WE films. Furthermore, a shorter τ_e results in higher photoelectron collection efficiency at the FTO substrate and higher J_{sc} values of devices. These results are in good agreement with the current-voltage (J-V) results.

Figure 4a shows the absorption spectra of only NPs layer-coated FTO glass substrate with respect to wavelength. As can be observed, the absorption spectra of AuNPs-coated FTO glass substrate becomes almost zero within the range of wavelength from 300 to 800 nm. It was close to the absorption of FTO glass substrate. The results indicate that the plasmonic effect of AuNPs incorporated on FTO glass substrate could be neglected. However, the surface plasmonic absorption of AgNPs was clearly presented in about 400 nm. The absorption peak of the AgNPs/FTO glass was shifted compared to the AgNPs/FTO glass in other studies [6,21] due to the broad size distribution with large AgNPs. The radiant light can be enhanced after being effectively coupled with plasmon absorption in the range of 300–600 nm due to the increase of the optical density around AgNPs [9]. According to that, there were more photons, which was usually absorbed by the dye molecules placed in the vicinity of AgNPs [21–23], resulting in the improvement of the device performance.

In order to investigate the effect of plasmonic structure on the enhancement of the photocurrent density of devices, we conducted IPCE measurements. The results are given in Figure 4b. As can be seen, the IPCE of the DSC fabricated with the FTO/Ag-NPs/TiO$_2$ WE was higher than those of the DSC based on the FTO/TiO$_2$ WE and the FTO/Au-NPs/TiO$_2$ WE over the visible wavelength range. The current value calculated from the overlap integral of the IPCE spectrum agrees well with the J_{sc} derived from the J–V characteristics. It is well-known that the IPCE is composed of the light-harvesting and charge collection efficiencies, the efficiency of the electron injection from the excited dye into the TiO$_2$ and the efficiency for the dye regeneration. In this study, identical compounds were used to fabricate all devices, except the WEs. Therefore, the effect of the efficiency for the dye regeneration was excluded in this work. It is reported that the plasmonic enhancement is estimated through the improvement of optical density near the metal surface. Thus, more photons could be harvested by dye located in the vicinity of the NPs, resulting in the increase of IPCE [21–23]. As we mentioned above, the plasmon effect of AuNPs/FTO glass substrate was not significant. Thus, IPCE of DSC fabricated on AuNPs/FTO was similar to that of DSC assembled on FTO glass substrate. In contrast, the IPCE of the device based on AgNPs/FTO glass substrate became higher than those of other two DSCs due to the transverse plasmon absorption at around 400 nm. The result is in good accordance to Ultraviolet–visible spectroscopy (UV-Vis) data and *J–V* results.

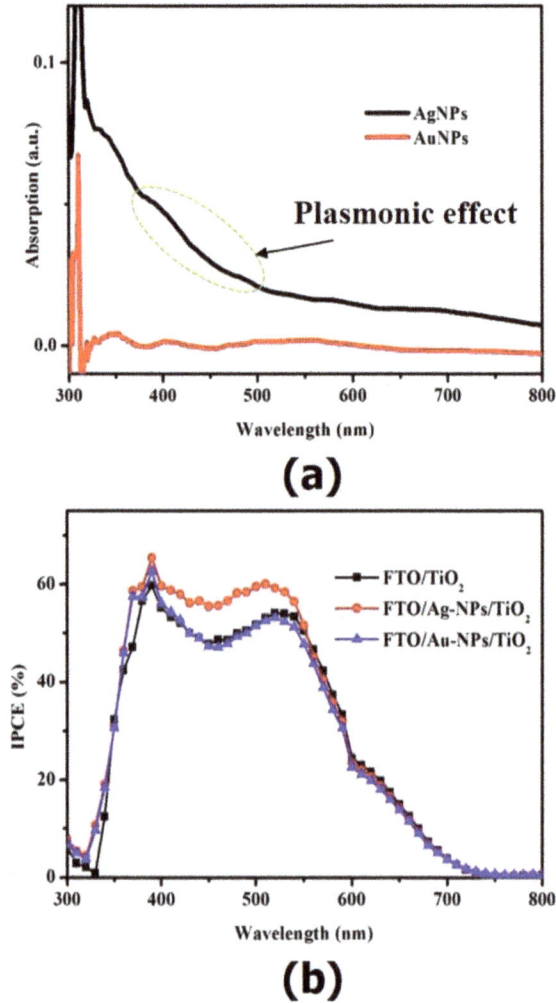

Figure 4. (a) Absorption spectra of AuNPs and AgNPs; (b) Incident photon-to-current efficiency (IPCE) spectra of DSCs based on different working electrodes. a.u.: arbitrary unit.

3. Materials and Methods

3.1. Materials

$H_2PtCl_6 \cdot xH_2O$ ($\geqslant 37.5\%$ Pt basic), $AgNO_3$ (>99%) and iso-propyl alcohol (IPA) (99.5%) were obtained from Sigma-Aldrich (St. Louis, MO, USA). $HAuCl_4 \cdot 3H_2O$ was purchased from Aldrich. FTO glass as a conductive transparency electrode was purchased from Solaronix, Aubonne, Switzerland (~8 Ω/square). These substrates were used after cleaning them by sonication in acetone (Fluka). The nonporous TiO_2 paste and ruthenium-based dye (N719) used in the study were purchased from

Solaronix, Aubonne, Switzerland. The dye was adsorbed from a 0.3 mM solution in a mixed solvent of acetonitrile (Sigma-Aldrich) and tert-butyl alcohol (Sigma-Aldrich) (St. Louis, MO, USA) with a volume ratio of 1:1. The electrolyte was a solution of 0.60 M 1-methyl-3-butylimidazolium iodide (Sigma-Aldrich), 0.03 M I_2 (Sigma-Aldrich), 0.10 M guanidinium thiocyanate (Sigma-Aldrich) and 0.50 M 4-tert-butylpyridine (Sigma-Aldrich) in a mixed solvent of acetonitrile (Sigma-Aldrich) and valeronitrile, with a volume ratio of 85:15.

3.2. Synthesis of Au and Ag on FTO Glass Substrates

Two solutions containing 10 mM $HAuCl_4 \cdot 3H_2O$ in IPA and 10 mM $AgNO_3$ in IPA were initially prepared. Separately, 8 µL of a precursor solution were deposited onto 2×2 cm^2 specimens of FTO glass, and the solvent was allowed to evaporate at 70 °C for 10 min. The specimens were then reduced using Ar plasma under atmospheric pressure at a power of 150 W, a gas flow rate of 5 litter per minute (lpm), a treatment time of 15 min and a substrate moving speed of 5 mm/s [10]. The morphologies of the Ag-NPs and Au-NPs on the FTO glass substrates were observed by high-resolution scanning electron microscopy (HRSEM). For the TEM measurements, we separately synthesized Ag-NPs and Au-NPs on Cu grids under the same plasma conditions used earlier.

3.3. Preparation of the Working Electrodes

Three WEs were prepared on FTO, Ag-NPs/FTO and Au-NPs/FTO glass substrates. The area of TiO_2 mesoporous layer and that of the TCO were 0.7×0.7 cm^2 and 2×2 cm^2, respectively. These WEs were fabricated through the following procedure. A transparent film of 20-nm TiO_2 particles (Solaronix, Switzerland) was coated onto the substrates by screen printing (200 T mesh), kept in a clean box for 3 min so that the paste could relax to reduce surface irregularities and then dried for 3 min at 125 °C. This screen-printing procedure (coating, storing and drying) was repeated until the thickness of the working electrode was approximately 12 µm. The electrodes coated with the TiO_2 paste were gradually heated under an air flow at 325 °C for 5 min, at 375 °C for 5 min, at 450 °C for 15 min and finally at 500 °C for 15 min under ambient conditions [10]. After cooling to 80 °C, the TiO_2 electrodes were immersed in a 0.3 mM Di-tetrabutylammonium cis-bis(isothiocyanato)bis(2,2'-bipyridyl-4,4'-dicarboxylato)ruthenium(II) (N719) dye solution in a mixture of acetonitrile (Sigma-Aldrich) and tert-butyl alcohol (Aldrich) (volume ratio of 1:1) and kept at room temperature for 24 h to complete the sensitizer uptake process.

3.4. Preparation of the Counter Electrodes

Pt CEs were also prepared through DPR, as described in our previous study [10].

3.5. Assembly and Measurement of the DSCs

The assembly and characterization of the DSCs were carried out as described in our previous study [10].

4. Conclusions

Au-NPs and Ag-NPs were successfully immobilized on FTO glass substrates by means of DPR, with the results showing homogeneous size dispersity. HRSEM and TEM images showed that the areal number density of Au-NPs/FTO was higher than that of Ag-NPs/FTO glass, while Au-NPs/FTO glass was smaller than that of Au-NPs/FTO glass. The DSC with FTO/Ag-NPs/TiO_2 as the WE demonstrated energy conversion efficiency of 7.49 ($\pm$0.14)%, which was better than those of DSCs with FTO/TiO_2 and FTO/Ag-NPs/TiO_2 WEs, *viz.* 6.54 ($\pm$0.15)% and 6.27 ($\pm$0.11)%, respectively.

Supplementary Materials: The following are available online at http://www.mdpi.com/2079-4991/6/4/70/s1.

Acknowledgments: This research was supported by a National Research Foundation (NRF) grant (2014R1A2A2A 01006994), through the Korea Research Fellowship Program (2015H1D3A1061830) and by a Korea Carbon Capture and Sequestration (CCS) R&D Center (KCRC) grant (2014M1A8A1049345). These were all funded by the Ministry of Science, ICT and Future Planning through the National Research Foundation of Korea.

Author Contributions: All authors planned the experiment and discussed the data. Van-Duong Dao prepared the manuscript, and Ho-Suk Choi reviewed it.

Conflicts of Interest: The authors declare no conflict of interest.

Abbreviations

The following abbreviations are used in this manuscript:

DSC	dye-sensitized solar cell
NPs	nanoparticles
FTO	fluorine-doped tin oxide
TCO	transparent conducting oxide
CE	counter electrode
WE	working electrode
DPR	dry plasma reduction
IPA	iso-propyl alcohol
EIS	electrochemical impedance spectroscopy
IPCE	incident photon-to-current efficiency

References

1. O'Regan, B.; Gratzel, M. A Low-Cost, High-Efficiency Solar Cell Based on Dye-Sensitized Colloidal TiO$_2$ Films. *Nature* **1991**, *353*, 737–740. [CrossRef]
2. Gratzel, M. Conversion of Sunlight to Electric Power by Nanocrystalline Dye-Sensitized Solar Cells. *J. Photochem. Photobiol. A* **2004**, *164*, 3–14. [CrossRef]
3. Ito, S.; Zakeeruddin, S.M.; Humphry-Baker, R.; Liska, P.; Charvet, R.; Comte, P.; Nazeeruddin, M.K.; Péchy, P.; Takata, M.; Miura, H.; *et al.* High-Efficiency Organic-Dye-Sensitized Solar Cells Controlled by Nanocrystalline-TiO$_2$ Electrode Thickness. *Adv. Mater.* **2006**, *18*. [CrossRef]
4. Hagfeldt, A.; Boschloo, G.; Sun, L.; Kloo, L.; Pettersson, H. Dye-Sensitized Solar Cells. *Chem. Rev.* **2010**, *110*. [CrossRef] [PubMed]
5. Ito, S.; Murakami, T.N.; Comte, P.; Liska, P.; Gratzel, C.; Nazeeruddin, M.K.; Gratzel, M. Fabrication of Thin Film Dye Sensitized Solar Cells with Solar to Electric Power Conversion Efficiency over 10%. *Thin Solid Films* **2008**, *516*, 4613–4619. [CrossRef]
6. Lin, S.-J.; Lee, K.-C.; Wu, J.-L.; Wu, J.-Y. Plasmon-Enhanced Photocurrent in Dye-Sensitized Solar Cells. *Solar Energy* **2012**, *86*, 2600–2605. [CrossRef]
7. Standridge, S.D.; Schatz, G.C.; Hupp, J.T. Distance Dependence of Plasmon-Enhanced Photocurrent in Dye-Sensitized Solar Cells. *J. Am. Chem. Soc.* **2009**, *131*, 8407–8409. [CrossRef] [PubMed]
8. Zhang, D.; Wang, M.; Brolo, A.G.; Shen, J.; Li, X.; Huang, S. Enhanced Performance of Dye-Sensitized Solar Cells Using Gold Nanoparticles Modified Fluorine Tin Oxide Electrodes. *J. Phys. D* **2013**, *46*. [CrossRef]
9. Chang, S.; Li, Q.; Xiao, X.; Wong, K.Y.; Chen, T. Enhancement of Low Energy Sunlight Harvesting in Dye-Sensitized Solar Cells Using Plasmonic Gold Nanorods. *Energy Environ. Sci.* **2012**, *5*, 9444–9448. [CrossRef]
10. Dao, V.D.; Tran, Q.C.; Ko, S.H.; Choi, H.S. Dry Plasma Reduction to Synthesize Supported Platinum Nanoparticles for Flexible Dye-Sensitized Solar Cells. *J. Mater. Chem. A* **2013**, *1*, 4436–4443. [CrossRef]
11. Dao, V.D.; Larina, L.L.; Lee, J.K.; Jung, K.D.; Huy, B.T.; Choi, H.S. Graphene-Based RuO$_2$ Nanohybrid as a Highly Efficient Catalyst for Triiodide Reduction in Dye-Sensitized Solar Cells. *Carbon* **2014**, *81*, 710–719. [CrossRef]
12. Dao, V.D.; Choi, Y.; Yong, K.; Larina, L.; Shevaleevskiy, O.; Choi, H.S. A Facile Synthesis of Bimetallic AuPt Nanoparticles as a New Transparent Counter Electrode for Quantum-Dot-Sensitized Solar Cells. *J. Power Sources* **2015**, *274*, 831–838. [CrossRef]
13. Ke, X.; Zhang, X.; Zhao, J.; Sarina, S.; Barry, J.; Zhu, H. Selective Reductions Using Visible Light Photocatalysts of Supported Gold Nanoparticles. *Green Chem.* **2013**, *15*, 236–244. [CrossRef]

14. Zheng, L.; Zhang, G.; Zhang, M.; Guo, S.; Liu, Z.H. Preparation and Capacitance Performance of Ag-Graphene Based Nanocomposite. *J. Power Sources* **2012**, *201*, 376–381. [CrossRef]
15. Jakob, M.; Levanon, H.; Kamat, P.V. Charge Distribution between UV-Irradiated TiO$_2$ and Gold Nanoparticles: Determination of Shift in the Fermi Level. *Nano Lett.* **2003**, *3*, 353–358. [CrossRef]
16. Martinson, A.B.F.; Hamann, T.W.; Pellin, M.J. New Architectures for Dye-Sensitized Solar Cells. *Chemistry* **2008**, *14*, 4458–4467. [CrossRef] [PubMed]
17. Dao, V.D.; Choi, H.S.; Jung, K.D. Effect of Ohmic Serial Resistance on the Efficiency of Dye-Sensitized Solar Cells. *Mater. Lett.* **2013**, *92*, 11–13. [CrossRef]
18. Gao, Z.; Wu, Z.; Li, X.; Chang, J.; Wu, D.; Ma, P.; Xu, F.; Gao, S.; Jiang, K. Application of Hierarchical TiO$_2$ Spheres as Scattering Layer for Enhanced Photovoltaic Performance in Dye-Sensitized Solar Cell. *CrystEngComm* **2013**, *5*, 3351–3358. [CrossRef]
19. Kern, R.; Sastrawan, R.; Ferber, J.; Stangl, R.; Luther, J. Modeling and Interpretation of Electrical Impedance Spectra of Dye Solar Cells Operated Under Open-Circuit Conditions. *Electrochim. Acta* **2002**, *47*, 4213–4225. [CrossRef]
20. Park, J.T.; Roh, D.K.; Patel, R.; Kim, E.; Ryu, D.Y.; Kim, J.H. Preparation of TiO$_2$ Spheres with Hierarchical Pores via Grafting Polymerization and Sol-Gel Process for Dye-Sensitized Solar Cells. *J. Mater. Chem.* **2010**, *20*, 8521–8530. [CrossRef]
21. Lin, S.J.; Lee, K.C.; Wu, J.L.; Wu, J.Y. Enhanced Performance of Dye-Sensitized Solar Cells via Plasmonic Sandwiched Structure. *Appl. Phys. Lett.* **2011**, *99*. [CrossRef]
22. Qi, J.; Dang, X.; Hammond, P.T.; belcher, A.M. Highly Efficient Plasmon-Enhanced Dye-Sensitized Solar Cells through Metal@Oxide Core–Shell Nanostructure. *ACS Nano* **2012**, *5*, 7108–7116. [CrossRef] [PubMed]
23. Ding, B.; Lee, B.J.; Yang, M.; Jung, H.S.; Lee, J.K. Surface-Plasmon Assisted Energy Conversion in Dye-Sensitized Solar Cells. *Adv. Energy Mater.* **2011**, *1*. [CrossRef]

nanomaterials

MDPI

Article

Synthesis of Lithium Metal Oxide Nanoparticles by Induction Thermal Plasmas

Manabu Tanaka [1], Takuya Kageyama [2], Hirotaka Sone [2], Shuhei Yoshida [3], Daisuke Okamoto [2] and Takayuki Watanabe [1,*]

[1] Department of Chemical Engineering, Faculty of Engineering, Kyushu University, 744 Motooka, Nishi-ku, Fukuoka 819-0395, Japan; mtanaka@chem-eng.kyushu-u.ac.jp

[2] Department of Chemical Systems and Engineering, Graduate School of Engineering, Kyushu University, 744 Motooka, Nishi-ku, Fukuoka 819-0395, Japan; kageyama.takuya.520@kyudai.jp (T.K.); h.sone@kyudai.jp (H.S.); daisuke.okamoto@kyudai.jp (D.O.)

[3] Chemical Engineering Course, School of Engineering, Kyushu University, 744 Motooka, Nishi-ku, Fukuoka 819-0395, Japan; s.yoshida732@kyudai.jp

* Correspondence: watanabe@chem-eng.kyusu-u.ac.jp; Tel./Fax: +81-92-802-2745

Academic Editors: Krasimir Vasilev and Melanie Ramiasa
Received: 25 January 2016; Accepted: 29 March 2016; Published: 6 April 2016

Abstract: Lithium metal oxide nanoparticles were synthesized by induction thermal plasma. Four different systems—Li–Mn, Li–Cr, Li–Co, and Li–Ni—were compared to understand formation mechanism of Li–Me oxide nanoparticles in thermal plasma process. Analyses of X-ray diffractometry and electron microscopy showed that Li–Me oxide nanoparticles were successfully synthesized in Li–Mn, Li–Cr, and Li–Co systems. Spinel structured $LiMn_2O_4$ with truncated octahedral shape was formed. Layer structured $LiCrO_2$ or $LiCoO_2$ nanoparticles with polyhedral shapes were also synthesized in Li–Cr or Li–Co systems. By contrast, Li–Ni oxide nanoparticles were not synthesized in the Li–Ni system. Nucleation temperatures of each metal in the considered system were evaluated. The relationship between the nucleation temperature and melting and boiling points suggests that the melting points of metal oxides have a strong influence on the formation of lithium metal oxide nanoparticles. A lower melting temperature leads to a longer reaction time, resulting in a higher fraction of the lithium metal oxide nanoparticles in the prepared nanoparticles.

Keywords: thermal plasmas; lithium metal oxide; nanoparticle formation mechanism

1. Introduction

Nanoparticles have become widely utilized due to their enhanced and unique properties relative to bulk materials. The preparation of nanoparticles can be classified into physical and chemical methods. The physical methods include mechanical milling [1,2], laser ablation [3–5], and other aerosol processes with energy sources to provide a high temperature. Among these methods, attractive material processing with thermal plasmas have been proposed for the nanoparticles production. This is because thermal plasmas offer unique advantages; high enthalpy to enhance reaction kinetics, high chemical reactivity, rapid quenching rate in the range of 10^{3-6} K/s, and selectivity of atmosphere in accordance with the required chemical reactions. Thermal plasmas are capable of evaporating large amount of raw materials, even with high melting and boiling temperatures [6–9]. Furthermore, high-purity nanoparticles can be synthesized in an induction thermal plasma because thermal plasma can be generated in a plasma torch without internal electrodes [10,11]. These advantages of thermal plasmas have brought about advances in plasma chemistry and plasma processing [12–15].

Lithium metal oxides have attracted many researchers because of their unique properties as cathode for lithium-ion batteries [16–18], CO_2 sorption material [19], and other magnetic, electrochemical materials. Layer structured $LiCoO_2$ is widely employed as cathodes in commercial battery applications, in spite of its toxicity and high cost. To solve the economic and environmental problems, alternatives of $LiCoO_2$ have been intensively explored. One of the alternatives is spinel-structured $LiMn_2O_4$, which provides a promising high-voltage cathode material for lithium-ion batteries owing to their high theoretical energy density, low cost, and eco-friendliness [18,20–22]. $LiNiO_2$ is also a candidate due to its excellent cycle life, with negligible capacity fading *etc.* [17,23]. The liquid phase method is generally used in the synthesis of lithium metal oxide nanoparticles; however, productivity of the nanoparticles in the liquid phase method is insufficient for industrial application. Therefore, the synthesis method of lithium oxide nanoparticles with a high productivity is strongly demanded.

Thermal plasmas are expected to be promising energy sources to fabricate nanoparticles at high productivity from micron-sized powder as raw materials. Here, one-step synthesis of lithium-metal oxide nanoparticles with induction thermal plasma is studied. The purpose of the present study is to synthesize the lithium metal oxide nanoparticles via the induction thermal plasma and to investigate the formation mechanism of lithium-metal oxide nanoparticles. Different lithium metal systems—Li–Mn, Li–Cr, Li–Co, and Li–Ni—were compared to understand the formation mechanism.

2. Results and Discussion

2.1. Experimental Results

Figure 1 shows X-ray diffraction spectra of the prepared nanoparticles by the induction thermal plasma in different systems—Li–Mn, Li–Co, Li–Cr, and Li–Ni. In the case of the Li–Mn system, main diffraction peaks correspond to spinel-structured $LiMn_2O_4$, while diffraction peaks of Mn_3O_4 are also found. In cases of Li–Co and Li–Cr, layer-structured $LiCoO_2$ and $LiCrO_2$ were found as well as their oxides. In the Li–Ni case, $Li_{0.4}Ni_{1.6}O_2$ and unreacted Li_2CO_3 were confirmed. These results clearly show the strong influence of the constituent metals on the formation of lithium oxide nanoparticles. Moreover, Figure 2 indicates composition of the prepared nanoparticles analyzed from X-ray diffractometry (XRD) spectra. These compositions of the prepared nanoparticles were used to evaluate the ratio of metal that reacted with Li. Evaluated reaction ratio will be discussed in the following section.

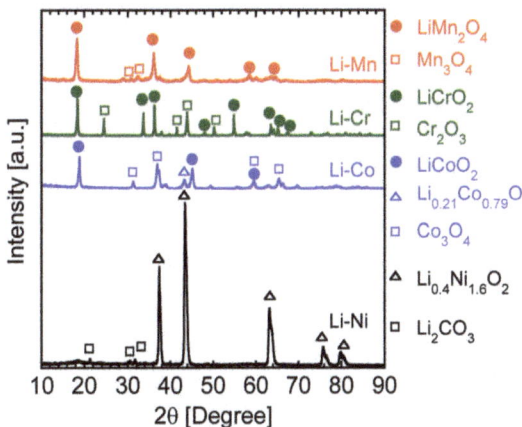

Figure 1. X-ray diffractometry (XRD) spectra of as-prepared nanoparticles by induction thermal plasma in different systems, Li–Mn, Li–Co, Li–Cr, and Li–Ni.

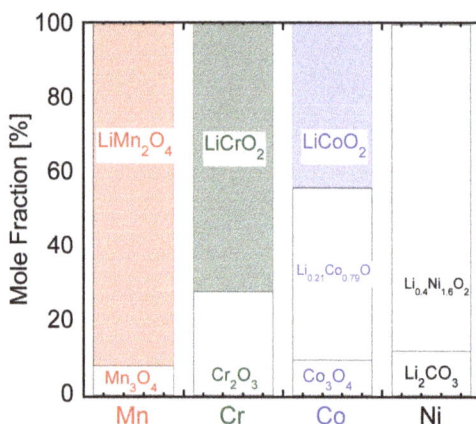

Figure 2. Composition of prepared nanoparticles by induction thermal plasma in different systems, Li–Mn, Li–Co, Li–Cr, and Li–Ni.

Morphologies and particle sizes of the prepared nanoparticles were observed via transmission electron microscopy (TEM). Figure 3 shows the representative TEM images and the particle size distributions of the nanoparticles in different systems. Many spherical particles with 60 nm in mean diameter are observed in the Li–Ni system. In the cases of Li–Mn, Li–Co, and Li–Cr, most of the particles with 50–80 nm in mean diameters have polyhedral shapes including quadrangular, pentagonal, and hexagonal shapes. In particular, many particles with a hexagonal shape can be found as shown in Figure 3a. Hence, the spinel-structured $LiMn_2O_4$ is considered to have a hexagonal shape because $LiMn_2O_4$ is a major product in the Li–Mn system, according to XRD results.

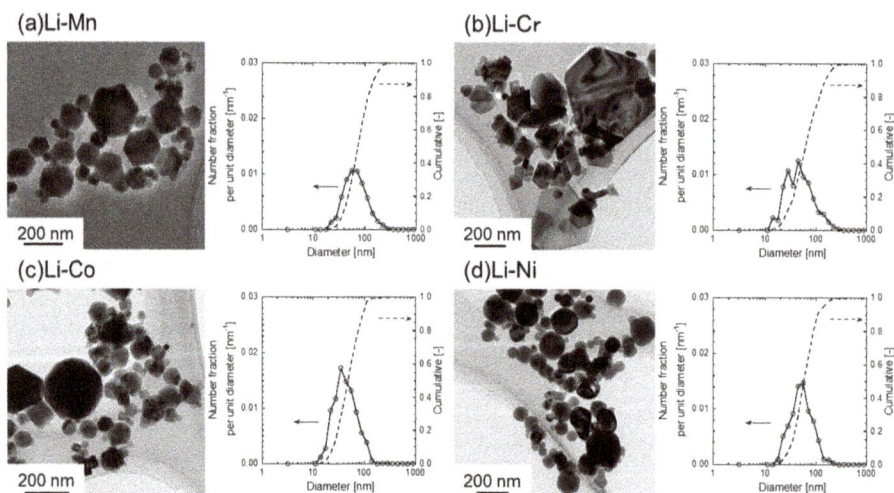

Figure 3. (**a**) Representative transmission electron microscopy (TEM) image and particle size distribution of prepared nanoparticles by induction thermal plasma in Li–Mn system; (**b**) that in Li–Cr system; (**c**) that in Li–Co system; (**d**) that in Li–Ni system.

Scanning electron microscopy (SEM) observation was carried out to clarify morphology of $LiMn_2O_4$ more specifically than TEM observation. Figure 4 shows the representative SEM image of the prepared nanoparticles in the Li–Mn system. Many particles have truncated an octahedral shape as shown in Figure 4. Previous research on morphology of ferrite nanoparticles synthesized by the induction thermal plasma reported that the spinel-structured nanoparticles had a truncated octahedral shape [24]. The spinel-structured nanoparticles synthesized in thermal plasma have a truncated octahedral shape, although the stable structure of the spinel-structured particles are generally considered to be of an octahedral shape. The reason for this truncation is currently under investigation.

Figure 4. Representative scanning electron microscopy (SEM) image of prepared nanoparticles in Li–Mn system.

2.2. Nanoparticle Formation Mechanism

Homogeneous nucleation temperatures of metals considered in the present study were estimated based on nucleation theory considering non-dimensional surface tension [25]. The homogeneous nucleation rate J can be expressed as

$$J = \frac{\beta_{ij} n_s^2 S}{12} \sqrt{\frac{\Theta}{2\pi}} \exp\left(\Theta - \frac{4\Theta^3}{27\left(\ln S\right)^2}\right) \qquad (1)$$

where S is the saturation ratio and n_s is the equilibrium saturation monomer concentration at temperature T. β is the collision frequency function. The dimensionless surface tension is given by the following equation:

$$\Theta = \frac{\sigma s_1}{kT} \qquad (2)$$

where σ is the surface tension and s_1 is the monomer surface area. The surface tension and the saturation ratio have a dominant influence on determining the nucleation rate. Stable nuclei are observed experimentally when the nucleation rate is over $1.0 \text{ cm}^{-3} \cdot \text{s}^{-1}$. Hence, the corresponding value of the saturation ratio is defined as the critical saturation ratio. The detailed procedure to estimate the nucleation rate and corresponding nucleation temperature can be found in previous works [10,25].

The relationship between the calculated nucleation temperature and the boiling and melting points is summarized in Figure 5. Because of the unknown properties of metal oxides, only melting point oxides are plotted for metal oxides. These temperatures indicate that the melting points of the oxides are higher than the nucleation temperatures of pure metals in each Li–Me system. Therefore, nucleation of metal oxides is considered to occur at first. Li oxide and metal vapors co-condense onto the nuclei just after the nucleation starts.

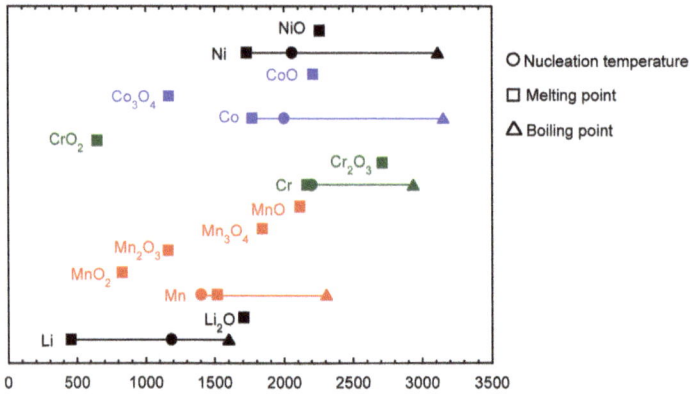

Figure 5. Relationship between nucleation temperature and melting and boiling points of considered metals and metal oxides.

The above mechanism can be proposed as a common mechanism for all considered Li–Me systems because the relationship between the nucleation temperature and the melting and boiling temperature of their oxides shows the same trend. However, experimental results show a different ratio of metal reactive with Li to total metal. In the cases of Li–Cr and Li–Mn, high ratios of the reactive metals were obtained, while a low ratio was obtained in Li–Ni system. Then, melting points of each oxide in different system were focused because the reaction rate of metal oxide particles and the condensed lithium would drastically decrease after complete solidification of the growing particles.

Figure 6 shows the relationship between the lowest melting temperature of metal oxides for each Li–Me system and the reaction ratios. The results suggest a lower melting point of oxide leads to a higher reaction ratio. This can be explained by the different reaction time during the nanoparticle formation process. Figure 7 summarizes the formation mechanism of Li–Me oxide nanoparticles. A lower melting point of oxide leads to a longer residence time of growing particles in a liquid-like state, resulting in the longer reaction time with condensed lithium oxide. Consequently, a higher reaction ratio can be obtained in the Li–Me system, as in that of Li–Mn and Li–Cr. On the other hand, a shorter residence time of the growing particles in a liquid-like state leads to a shorter reaction time, resulting in a lower reaction ratio. These results suggest that the melting point of metal oxide has a strong influence on the reaction with lithium oxide.

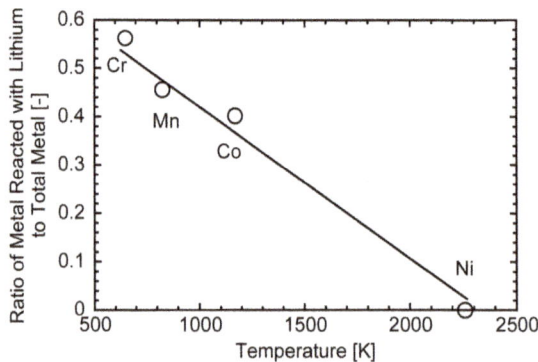

Figure 6. Relationship between lowest melting points of metal oxides and the reaction ratios for each Li–Me system.

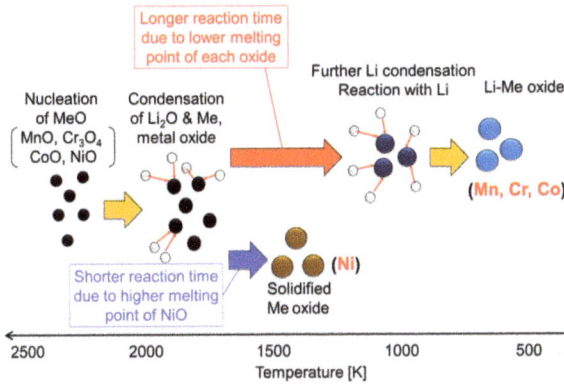

Figure 7. Conceptual diagram of formation mechanism of Li–Me oxide nanoparticles in induction thermal plasma.

The above mechanism can explain the obtained results well, although this is still only a hypothesis. Further experimental and theoretical investigation will be required to understand the formation mechanism of complicated oxide nanoparticles in thermal plasmas.

3. Experimental Section

3.1. Experimental Setup and Conditions

A schematic illustration of the experimental setup for Li–Me oxide nanoparticle fabrication is presented in Figure 8a. This equipment is composed of the plasma torch, the reaction chamber where the nanoparticles are synthesized, and the filter unit. Figure 8b shows the enlarged illustration of the plasma torch. High temperature plasma of more than 10,000 K was generated in the plasma torch by induction heating at 4 MHz. The input power was controlled at 20 kW. Ar was used for the carrier gas of the raw powder at 3 L/min and inner gas at 5 L/min. A mixture of argon and oxygen were used for the plasma forming gas at 60 L/min. Mixture of Li_2CO_3 with 3.5 μm in diameter and metal or metal oxide with 3–10 μm in mean diameter were injected from the powder feeder by Ar carrier gas. Different metals including Mn, Cr, Co, and Ni were compared to investigate the formation mechanism of Li–Me (Mn, Cr, Co, and Ni) nanoparticles. The composition ratio of Li_2CO_3 to metal was 0.5. The powder feed rate was fixed at 400 mg/min. These experimental conditions are listed in Table 1.

Prepared nanoparticles were collected from the filter and the inner wall of the reaction chamber. Unevaporated raw materials were not confirmed according to SEM observation of the collected particles. This fact implies that the fed raw materials were completely evaporated in the high temperature region of the thermal plasma and converted into nanoparticles during the quenching process. Therefore, the mole fractions of the Li–Me oxides indicated in Figure 2 correspond to the yields of the Li–Me oxide in this nanoparticle fabrication process by the induction thermal plasma.

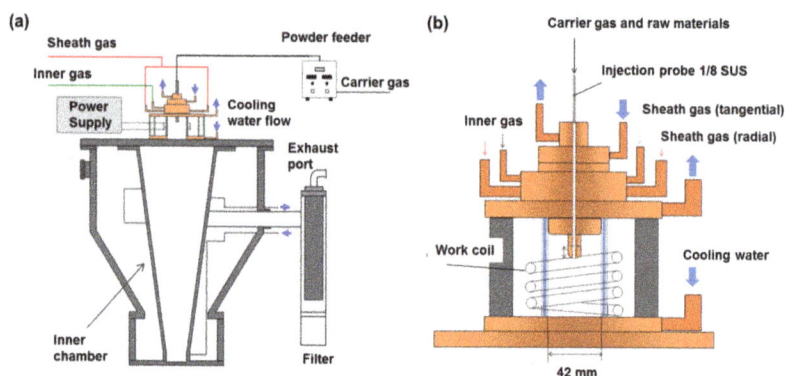

Figure 8. Schematic illustration of induction thermal plasma system for nanoparticles fabrication (**a**) and enlarged illustration of plasma torch (**b**).

Table 1. Typical experimental conditions.

Plasma Conditions			
Input power		20 kW	
Frequency		4 MHz	
Pressure		101.3 kPa	
Sheath gas	Ar: 57.5 L/min		O_2: 2.5 L/min
Inner gas		Ar: 5 L/min	
Carrier gas		Ar: 3 L/min	
Discharge time		5 min	
Feed rate		400 mg/min	
Raw Materials			
System	Li–Mn	Li–Cr	Li–Co	Li–Ni
Raw powders	Li_2CO_3, MnO_2	Li_2CO_3, Cr	Li_2CO_3, Co	Li_2CO_3, Ni
Li/Me ratio			0.5	

3.2. Characterization of Prepared Nanoparticles

The phase identification of the prepared nanoparticles was determined by X-ray diffractometry (XRD, Multiflex, Rigaku Co., Tokyo, Japan), operated with Cu Kα source (λ = 0.1541 nm). The diffraction data was collected using a continuous scan mode with a speed of 2 degree/min in the region of 10–90 degrees with a step size of 0.04 degrees. The accelerating voltage and applied current was 40 kV and 50 mA, respectively. The quantitative analysis of the composition of the prepared nanoparticles was conducted based on the whole-powder-pattern-decomposition (WPPD) method with the assumption that no amorphous particles were included in the prepared nanoparticles.

The particle morphology and size distribution of the prepared nanoparticles were observed by TEM (JEM-2100HCKM, JEOL Ltd., Tokyo, Japan), operated at an accelerating voltage of 200 kV. The TEM specimens were prepared by dispersing the as-prepared nanoparticles in ethanol and placing a few drops of the dispersion on a carbon-grid. Furthermore, the 3D particle morphology was observed by field emission (FE)-SEM (SII TES+ Zeiss ULTRA55, Carl Zeiss, Oberkochen, Germany).

4. Conclusions

Lithium metal oxide nanoparticles were synthesized in induction thermal plasma and formation mechanism was investigated. Obtained remarks are as follows:

(a) Lithium metal oxide nanoparticles were synthesized in different Li–Me (Mn, Cr, Co, and Ni) systems. In the case of Li–Mn, Li–Cr, and Li–Co, lithium-metal oxide nanoparticles were successfully synthesized, while Li–Ni oxides were not synthesized in the Li–Ni system.

(b) The spinel-structured $LiMn_2O_4$ with a truncated octahedral shape was synthesized in Li–Mn system, although the stable shape of the spinel structure was an octahedral shape.

(c) The relationship between nucleation temperature and boiling and melting points of the considered metals and their oxides suggests the following formation mechanism: Metal oxide starts to nucleate at first. Then, vapors of metal and lithium oxide co-condense on the metal nuclei with an oxidation reaction.

(d) Melting point of metal oxides is an important factor in determining the final product of the Li–Me composite. A lower melting point of metal oxide leads to a longer reaction time, resulting in higher yields of the Li–Me composite.

(e) Nanomaterial fabrication with induction thermal plasma enables the production of high-purity nanoparticles of Li–Me oxide at high productivity.

Acknowledgments: The authors thank the Ultramicroscopy Research Center, Kyushu University for TEM and FE-SEM observation. The authors also thank the Center of Advanced Instrumental Analysis, Kyushu University for XRD analyses.

Author Contributions: Manabu Tanaka, Hirotaka Sone and Takayuki Watanabe coordinated the study. Manabu Tanaka, Takuya Kageyama, Shuhei Yoshida and Daisuke Okamoto conducted the experiments. Takuya Kageyama, Hirotaka Sone and Shuhei Yoshida conducted analyses of the produced nanoparticles. Takuya Kageyama conducted the data analyses. Manabu Tanaka and Takayuki Watanabe prepared the manuscript. All authors read and approved the manuscript.

Conflicts of Interest: The authors declare no conflict of interest.

1. Janot, R.; Guerard, D. One-step synthesis of maghemite nanometric powders by ball-milling. *J. Alloys Compd.* **2002**, *333*, 302–307. [CrossRef]
2. Pithawalla, Y.B.; El Shall, M.S.; Deevi, S.C. Synthesis and characterization of nanocrystalline iron aluminide particles. *Intermetallics* **2000**, *8*, 1225–1231. [CrossRef]
3. Peterson, S.; Barickowski, S. *In situ* bioconjugation: Single step approach to tailored nanoparticle-bioconjugates by ultrashort pulsed laser ablation. *Adv. Funct. Mater.* **2009**, *19*, 1167–1172. [CrossRef]
4. Mafune, F.; Khono, J.Y.; Takeda, Y.; Kondow, T. Formation of gold nanoparticles by laser ablation in aqueous solution of surfactant. *J. Phys. Chem. B* **2001**, *105*, 5114–5120. [CrossRef]
5. Yasukuni, R.; Horinaka, T.; Asahi, T. Preparation of perylenediimide nanoparticle colloids by laser ablation in water and their optical properties. *Jpn. J. Appl. Phys.* **2010**, *49*. [CrossRef]
6. Shigeta, M.; Watanabe, T. Numerical analysis for co-condensation processes in silicide nanoparticle synthesis using induction thermal plasmas at atmospheric pressure conditions. *J. Mater. Res.* **2005**, *20*, 2801–2811. [CrossRef]
7. Shigeta, M.; Watanabe, T. Growth mechanism of silicon-based functional nanoparticles fabricated by inductively coupled thermal plasmas. *J. Phys. D* **2007**, *40*, 2407–2419. [CrossRef]
8. Shigeta, M.; Watanabe, T. Numerical investigation of cooling effect on platinum nanoparticle formation in inductively coupled thermal plasmas. *J. Appl. Phys.* **2008**, *103*. [CrossRef]
9. Shigeta, M.; Murphy, A.B. Thermal plasmas for nanofabrication. *J. Phys. D* **2011**, *44*. [CrossRef]
10. Tanaka, M.; Noda, J.; Watanabe, T.; Matsuno, J.; Tsuchiyama, A. Formation mechanism of metal embedded amorphous silicate nanoparticles by induction thermal plasmas. *J. Phys. Conf. Ser.* **2014**, *518*. [CrossRef]
11. Cheng, Y.; Tanaka, M.; Watanabe, T.; Choi, S.-Y.; Shin, M.-S.; Lee, K.-H. Synthesis of Ni_2B nanoparticles by RF thermal plasma for fuel cell catalyst. *J. Phys. Conf. Ser.* **2014**, *518*. [CrossRef]
12. Shigeta, M.; Watanabe, T. Effect of precursor fraction on silicide nanopowder growth under a thermal plasma condition: A computational study. *Powder Technol.* **2015**, *288*, 191–201. [CrossRef]
13. Watanabe, T.; Liu, Y.; Tanaka, M. Investigation of electrode phenomena in an innovative thermal plasma for glass melting. *Plasma Chem. Plasma Proc.* **2014**, *34*, 443–456. [CrossRef]

14. Liang, F.; Tanaka, M.; Watanabe, T. Measurement of anode surface temperature in carbon nanomaterial production by arc discharge method. *Mater. Res. Bull.* **2014**, *60*, 158–165. [CrossRef]

15. Tanaka, M.; Watanabe, T. Enhanced vaporization from molten metal surface by argon-hydrogen arc plasma. *Jpn. J. Appl. Phys.* **2013**, *52*. [CrossRef]

16. Xiao, X.; Wang, L.; Wang, D.; He, Q.; Peng, Q.; Li, Y. Hydrothermal synthesis of orthorhombic LiMnO$_2$ nano-particles and LiMnO$_2$ nanorods and comparison of their electrochemical performances. *Nano. Res.* **2009**, *2*, 923–930. [CrossRef]

17. Kalyani, P.; Kalaiselvi, N. Various aspects of LiNiO$_2$ chemistry: A review. *Sci. Technol. Adv. Mater.* **2005**, *6*, 689–703. [CrossRef]

18. Curtis, C.J.; Wang, J.; Schulz, D.L. Preparation and characterization of LiMn$_2$O$_4$ spinel nanoparticles as cathode materials in secondary Li batteries. *J. Electrochem. Soc.* **2004**, *151*, A590–A598. [CrossRef]

19. Ida, J.; Lin, Y.S. Mechanism of high-temperature CO$_2$ sorption on lithium zirconate. *Environ. Sci. Technol.* **2003**, *37*, 1999–2004. [CrossRef] [PubMed]

20. Shaju, K.M.; Bruce, P.G. Macroporous Li(Ni$_{1/3}$Co$_{1/3}$Mn$_{1/3}$)O$_2$: A high power and high energy cathode for rechargeable lithium batteries. *Adv. Mater.* **2006**, *18*. [CrossRef]

21. Canulescu, S.; Papadopoulou, E.L.; Anglos, D.; Lippert, T.; Schneider, W.; Wokaun, A. Mechanism of the laser plume expansion during the ablation of LiMn$_2$O$_4$. *J. Appl. Phys.* **2009**, *105*. [CrossRef]

22. Kim, D.K.; Muralidharan, P.; Lee, H.-W.; Ruffo, R.; Yang, Y.; Chan, C.K.; Peng, H.; Huggins, R.A.; Cui, Y. Spinel LiMn$_2$O$_4$ nanorods as lithium ion battery cathodes. *Nano Lett.* **2008**, *8*, 3948–3952. [CrossRef] [PubMed]

23. Ohzuku, T.; Ueda, A.; Nagayama, M. Electrochemistry and structural chemistry of LiNiO$_2$ (R3m) for 4 volt secondary lithium cells. *J. Electrochem. Soc.* **1993**, *140*, 1862–1870. [CrossRef]

24. Swaminathan, R.; Willard, M.A.; McHenry, M.E. Experimental observations and nucleation and growth theory of polyhedral magnetic ferrite nanoparticles synthesized using an RF plasma torch. *Acta Mater.* **2006**, *54*, 807–816. [CrossRef]

25. Girshick, S.L.; Chiu, C.-P.; McMurry, P.H. Time-dependent aerosol models and homogenous nucleation rates. *Aerosol Sci. Technol.* **1990**, *13*, 465–477. [CrossRef]

nanomaterials

MDPI

Article

Effect of Saturation Pressure Difference on Metal–Silicide Nanopowder Formation in Thermal Plasma Fabrication

Masaya Shigeta [1,*] and Takayuki Watanabe [2]

[1]   Joining and Welding Research Institute, Osaka University, 11-1 Mihogaoka, Ibaraki, Osaka 567-0047, Japan
[2]   Department of Chemical Engineering, Kyushu University, 744 Motooka, Nishi-ku, Fukuoka 819-0395, Japan;
      watanabe@chem-eng.kyushu-u.ac.jp
*   Correspondence: shigeta@jwri.osaka-u.ac.jp; Tel.: +81-6-6879-8648; Fax: +81-6-6879-8648

Academic Editors: Krasimir Vasilev and Thomas Nann
Received: 25 December 2015; Accepted: 1 March 2016; Published: 7 March 2016

Abstract: A computational investigation using a unique model and a solution algorithm was conducted, changing only the saturation pressure of one material artificially during nanopowder formation in thermal plasma fabrication, to highlight the effects of the saturation pressure difference between a metal and silicon. The model can not only express any profile of particle size–composition distribution for a metal–silicide nanopowder even with widely ranging sizes from sub-nanometers to a few hundred nanometers, but it can also simulate the entire growth process involving binary homogeneous nucleation, binary heterogeneous co-condensation, and coagulation among nanoparticles with different compositions. Greater differences in saturation pressures cause a greater time lag for co-condensation of two material vapors during the collective growth of the metal–silicide nanopowder. The greater time lag for co-condensation results in a wider range of composition of the mature nanopowder.

Keywords: nanopowder; metal silicide; co-condensation; thermal plasma; modelling

1. Introduction

Thermal plasmas have been used for effectual fabrication of nanopowders composed of nanometer-scale particles [1]. Nanopowders have unique capabilities that differ greatly from those of bulk materials or powders composed of larger particles [2]. Particularly, nanopowders composed of metal–silicide nanoparticles are anticipated to be potentially useful materials for extremely small electronic and mechanical applications such as solar-controlled windows, electromagnetic shielding, and contact materials in microelectronics [3]. However, because those raw materials usually have high melting points or boiling points, high-rate fabrication of those nanopowders is almost impossible using conventional methods such as grinding techniques and liquid-phase preparation. Combustion processes are also unusable because they are accompanied by unfavorable production of contaminants attributable to the oxidation atmosphere and because their flames cannot reach sufficiently high temperatures to vaporize the raw materials. Thermal plasmas offer the distinct benefits of high enthalpy, high chemical reactivity, variable properties, and a high cooling rate, all of which suit high-rate fabrication of metal–silicide nanopowders [4]. Additionally, the temperature and flow fields are controllable using external electromagnetic fields [5–7].

Thermal plasma fabrication of metal–silicide nanopowders involves the vaporization of raw materials and the subsequent conversion of the binary material vapors into numerous nanoparticles by virtue of the high enthalpy and high cooling rate of thermal plasma. However, the nanopowder growth is tremendously complicated because the nanopowder grows in a few tens of milliseconds through simultaneous and collective processes of binary homogeneous nucleation, binary heterogeneous co-condensation, and coagulation among nanoparticles with different compositions. Therefore, observing the growth process directly during experimentation is impossible. Only the characteristics of the final products have been evaluated [8–12]. Therefore, the growth mechanism remains poorly understood.

Computational studies based on theoretical modelling can reveal the growth mechanism and can enable prediction of the profile of the nanopowder to be synthesized. However, because of computational resource limitations, molecular dynamics (MD) calculation cannot comprehensively treat the entire growth process from nucleation until a nanopowder completes its growth [13]. In place of MD calculation with a heavy computational load, models based on aerosol dynamics have been used to simulate the process of a collective nanopowder growth comprehensively. Nevertheless, most models are applicable only to unary systems [14–21].

Only a few aerosol-dynamics-based models have been developed for thermal plasma fabrication of nanopowders involving co-condensations of binary material vapors [22–24]. Those models adopted several oversimplifications to obtain simple numerical solutions including only mean values. For more accurate and detailed numerical analysis in the growth processes of binary material nanopowders, we developed a unique model and solution algorithm [25–28]. That model can not only express any profile of particle size–composition distribution (PSCD) of a nanopowder even with widely ranging sizes from sub-nanometers to a few hundred nanometers, but can also simulate the entire formation process involving binary homogeneous nucleation, binary heterogeneous co-condensation, and coagulation among nanoparticles with different compositions. Especially for nanopowder formation of metal–silicides (Mo–Si, Ti–Si, Co–Si) under thermal plasma conditions, the model produced numerical results that agreed with experiment results [26,28].

Those computational results showed that the difference in saturation pressures between a metal and silicon was a crucial factor that determined the time lag of co-condensation and consequently affected the range of the silicon content in the synthesized nanopowder. Actually, an experimental study also reported that the saturation pressure difference affected the nanoparticle composition [11]. The literature emphasized that the difference of the saturation pressure caused that of the nucleation temperature and the larger difference resulted in a larger composition range of metal–silicide nanoparticles. Following this experimental study, the effect was investigated computationally using a simpler model as well [29]. Although the model did not consider binary nucleation and coagulation, it also predicted that systems with large differences of saturation pressures tended to produce metal–silicide nanoparticles with a wide range of compositions due to a time-lag of condensations of two materials.

Even though those studies indicated the importance of the saturation pressure difference between a metal and silicon, the effect remained unclear because the formation processes of nanopowders were also affected by material properties other than the saturation pressure. Data from the actual material properties were used for each material in those computations [26,28,29]. In experiments, it is generally impossible to control only a saturation pressure by changing materials. Other properties are changed as well. Therefore, in this study, numerical experiments are performed by changing only the saturation pressure of one material artificially to highlight the effects of saturation pressure differences on the metal–silicide nanopowder formation using the model which can simulate collective formation through simultaneous processes of binary nucleation, binary co-condensation, and coagulation among nanoparticles with different compositions [26,28].

2. Computational Conditions and Strategy

Figure 1 presents a schematic illustration of metal–silicide nanopowder fabrication using induction thermal plasma (ITP). The precursory raw materials are injected into a plasma where the high-temperature field vaporizes materials completely [22]. Metal and silicon vapors are transported with the flow to the plasma's tail, which exhibits a rapid temperature decrease.

Consequently, either or both of the material vapors become supersaturated, which engenders homogeneous nucleation. Because it is a binary system, nuclei composed of the metal atoms and silicon atoms are generated (binary nucleation). Immediately, the binary material vapors co-condense heterogeneously on the nuclei (binary co-condensation). Furthermore, during their growth, the nanoparticles mutually collide and merge into larger nanoparticles (coagulation). The metal–silicide nanopowder growth consists of these three processes that progress collectively and simultaneously. As a consequence, such nanopowders always have varieties of sizes and compositions, as shown in the experiments [11,22,28].

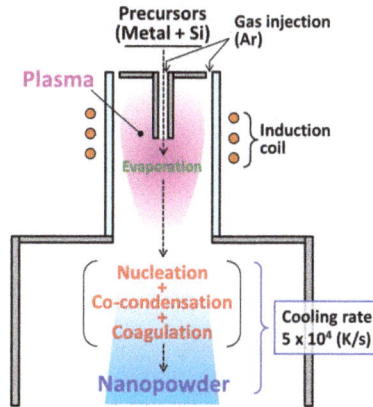

Figure 1. Metal–silicide nanopowder fabrication using an induction thermal plasma.

In a typical condition of ITP discharge, the region downstream from the plasma offers a high cooling rate of 10^4–10^5 K/s. Therefore, the present computation sets a constant cooling rate of 5.0×10^4 K/s to investigate the effect of saturation pressure difference under extremely simple conditions. The initial mole fraction of the material vapor to argon gas is set to be 0.5%. This can be regarded as a dilute condition in which the effect of the raw material on the flow field is negligible. This study particularly selects a titanium–silicon binary system because the saturation pressures of titanium and silicon are mutually close ($p_{S(\mathrm{Ti})}/p_{S(\mathrm{Si})} = 10^{-1}$–$10^0$). The initial ratio of these materials is set fairly at Ti:Si = 1:1. These conditions suggest that the vapors of titanium and silicon co-condense almost simultaneously.

As a great benefit of computational investigation, the value of only one saturation pressure can be changed strategically, which cannot be done in experiments. Therefore, this study defines an artificial saturation pressure $p_S' = \zeta \cdot p_S$. Controlling only the value of ζ from 10^{-3} to 10^3, while using the other actual material properties [30] with no changes, the effect of saturation pressure difference is highlighted. Computations are performed with a time increment Δt of 2.0 µs, which provides sufficient resolution for the present condition.

3. Outline of Metal–Silicide Nanopowder Formation Model

Formation of binary metal-silicide nanopowder from the vapor phase can be computed using a unique model developed by the authors [26–28]. The model with the PSCD describes a collective and simultaneous growth process of two-component nanoparticles in a binary vapor system through binary homogeneous nucleation, binary heterogeneous co-condensation, and coagulation among nanoparticles with different compositions. The PSCD is defined on a 2D coordinate system with two individual variables of the particle size and composition (here, silicon content), where nanoparticles composing a nanopowder are present only at the grid points [26].

In the model, the free energy of cluster formation W [31] is an important variable that dominates the nucleation and co-condensation:

$$W = -n_{(A)}k_B T \ln \left(\frac{N_{mono(A)}}{N'_{S(A)}} \right) - n_{(B)}k_B T \ln \left(\frac{N_{mono(B)}}{N'_{S(B)}} \right) + \sigma' s' \tag{1}$$

where $N_{mono(M)}$ stands for the monomer number density of material M (= A or B), $N'_{S(M)}$ signifies the equilibrium monomer number density of material M in the saturated vapor over a bulk solution. In addition, σ' and s' respectively represent the surface tension and the surface area of the cluster. For binary clusters, σ' is estimated approximately as:

$$\sigma' = \frac{n_{(A)}\sigma_{(A)} + n_{(B)}\sigma_{(B)}}{n_{(A)} + n_{(B)}} \tag{2}$$

Therein, $n_{(M)}$ is monomers of material M contained in a binary cluster and $\sigma_{(M)}$ is surface tension of material M. It is noteworthy that the nanoparticles are allowed to grow by condensation only when the free energy gradients for particle formation, W, is negative or zero. Therefore:

$$\frac{\partial W}{\partial n_{(M)}} \leqslant 0 \tag{3}$$

During a nanopowder growth process with a temperature decrease, the nanoparticles will be solidified. Then they can no longer increase their size as spherical particles by coagulation. In general, the solidification point depends on the material composition [32]. Furthermore, the solidification point decreases with the particle diameter [33]. Although these effects of the solidification point variation of binary-component nanoparticles were considered in our previous studies [26–28], the present study removes these effects to clarify only the effect of saturation pressure difference on the growth process.

4. Results and Discussion

Figure 2 shows the PSCD evolution of the Ti–Si nanopowder for $\zeta = 1$, which describes the growth process in an actual Ti–Si binary system. Figure 3a presents the histories of the vapor pressures and the saturation pressures of Ti and Si, whereas Figure 3b depicts the conversion ratios that indicate how much of each material vapor has been converted into nanoparticles. It is noted that the horizontal axes show the temperature in the opposite direction because of the cooling process. According to an earlier study [34], the nucleation rate with 1 nucleus/cm^{3}s can be conveniently observed experimentally. In this study, the nucleation is judged to start when the nucleation rate first exceeds this value. For the present case, nucleation starts when the vapors of Ti and Si are cooled to a temperature lower than 2444 K. Figure 2a shows that nuclei composed of Ti and Si are generated at the early stage of growth. At this time, the vapors of both Ti and Si are supersaturated, as shown in Figure 3a. Following nucleation, the nanoparticles grow rapidly (Figure 2b–e) by the simultaneous co-condensation of the material vapors on the nuclei as portrayed in Figure 3b as well as coagulation among themselves. Figure 3b also shows that 99% of Ti vapor completes the conversion at 2055 K, whereas 99% of Si vapor completes the conversion at 1977 K. After this drastic growth, the nanopowder grows slowly through coagulation and finally reaches its mature state of Figure 2f at 1716 K. To determine such a mature state of a nanopowder, the maximum difference of the particle number density at each node for Δt was monitored. That monitored parameter was defined as:

$$Q = \max \left(\frac{\left| \hat{N}_{i,j}^{(t)} - \hat{N}_{i,j}^{(t-\Delta t)} \right|}{\hat{N}_{i,j}^{(t-\Delta t)}} \right) \tag{4}$$

where

$$\hat{N}_{i,j}^{(t)} = \frac{N_{i,j}^{(t)}}{\rho_g^{(t)}} \tag{5}$$

and $\rho_g^{(t)}$ is the bulk gas density at the time t. When Q fell to less than 0.1, the nanopowder was determined to be mature. The mature nanopowder is composed mainly of the nanoparticles with the silicon content of $x_{(Si)}$ = 50.0 atom %, which is identical to the initially given silicon fraction to the precursor. Particle diameters range widely from a few nanometers to 76 nm.

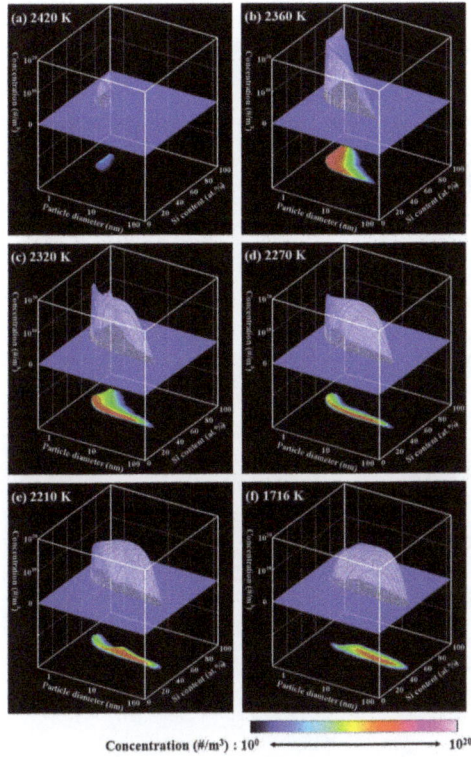

Figure 2. PSCD evolution for $\zeta = 1$.

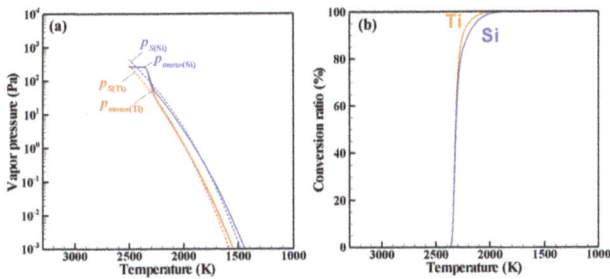

Figure 3. Phase conversion histories for $\zeta = 1$: (a) vapor pressures and (b) conversion ratios.

Figure 4 shows the PSCD evolution of the Ti–Si nanopowder for $\zeta = 10^3$. This value means that the saturation pressure of silicon is artificially set to be 1,000 times larger than the actual saturation pressure. Figure 5a,b respectively indicate the histories of the vapor pressures and the conversion ratios. Nucleation starts at 2258 K. At the early stage of growth, Ti-rich nuclei are generated (Figure 4a) and Ti-rich nanoparticles are formed (Figure 4b) only by Ti vapor condensation because the Si vapor pressure is still much lower than the saturation pressure (Figure 5a). During this growth of the Ti-rich nanoparticles, Si vapor starts to co-condense on the Ti-rich nanoparticles with Ti vapor (Figure 4c–e). Figure 5a shows that 99% of Ti vapor completes the conversion at 1882 K. After this co-condensation, Si vapor continues to condense on the Ti–Si nanoparticles slowly; 99% of Si vapor is consumed at 1506 K. The nanopowder reaches its mature state shown in Figure 4f at 1408 K. Although most the mature nanopowder exhibits silicon contents of $x_{(Si)} = 50.0$ atom %, the composition ranges widely from 25 atom % to 80 atom %.

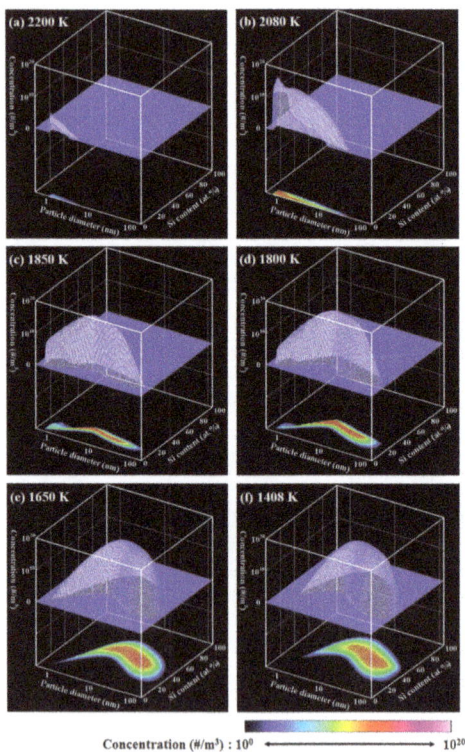

Figure 4. PSCD evolution for $\zeta = 10^3$.

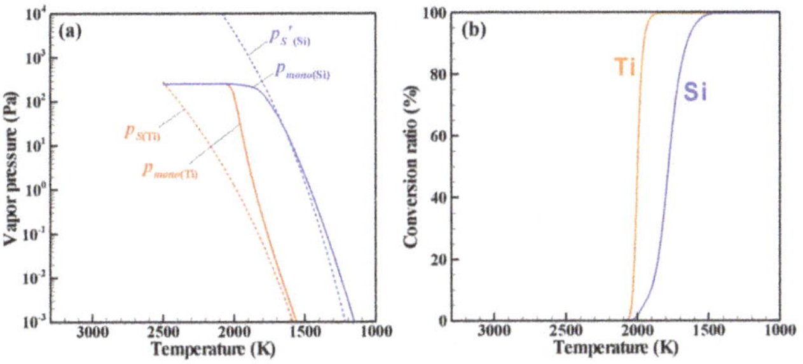

Figure 5. Phase conversion histories for $\zeta = 10^3$: (**a**) vapor pressures and (**b**) conversion ratios.

Figure 6 shows the PSCD evolution of the Ti–Si nanopowder for $\zeta = 10^{-3}$. Figure 7a,b respectively depict the vapor pressures and the conversion ratios. Nucleation starts at 3777 K, which is much higher than in the other cases. At the early stage of the growth, Si-rich nuclei are generated (Figure 6a) and Si-rich nanoparticles are formed (Figure 6b) by Si vapor condensation because only the Si vapor pressure is supersaturated (Figure 7a). Following this growth of Si-rich nanoparticles, Ti vapor starts to co-condense on the Si-rich nanoparticles with Ti vapor (Figure 6c–e). 99% of Si vapor is consumed at 2654 K, at which only 9% of Ti vapor is consumed (Figure 7b). 99% of Ti vapor completes the conversion at 2031 K. The nanopowder reaches its mature state as seen in Figure 6f at 1894 K. This mature nanopowder also has widely ranging silicon contents, from 25 atom % to 80 atom %.

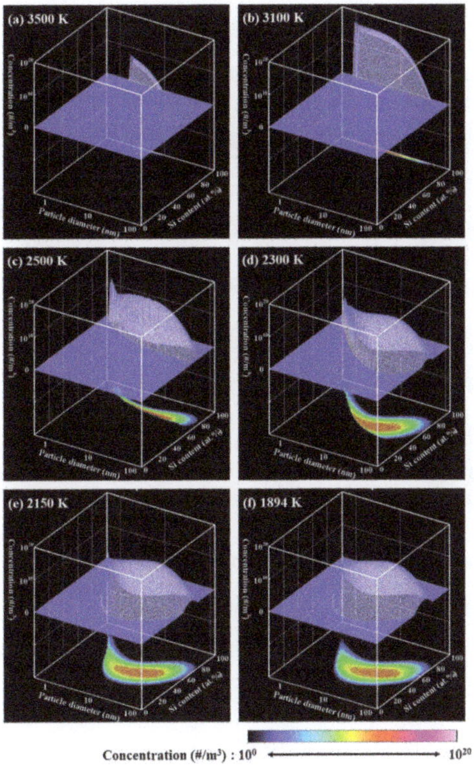

Figure 6. PSCD evolution for $\zeta = 10^{-3}$.

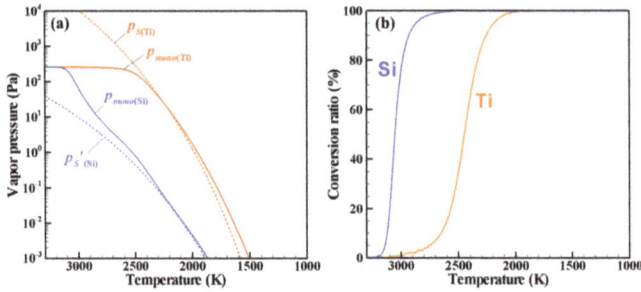

Figure 7. Phase conversion histories for $\zeta = 10^{-3}$: (**a**) vapor pressures and (**b**) conversion ratios.

These results imply that a wider range of composition is caused by a larger time lag of co-condensation that originally results from a larger difference of saturation pressures. Figure 8 shows the results obtained in the same manner using numerical experiments. As presumed, the standard deviations of fraction are larger when the saturation pressure differences are larger. Additionally, the results show that the standard deviations of size normalized by the arithmetic mean diameters are larger when the saturation pressure differences are larger.

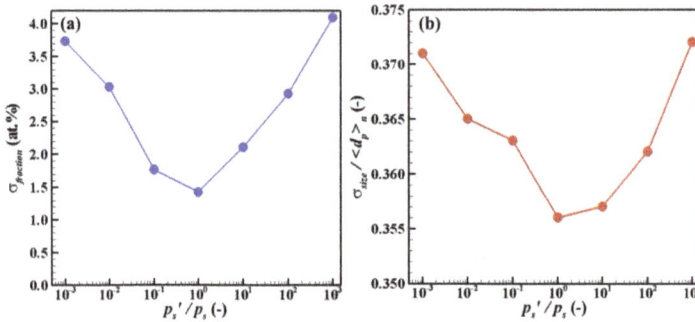

Figure 8. Effects of saturation pressure difference on dispersion of (**a**) fraction and (**b**) size.

5. Conclusions

Using our unique model and solution algorithm, which can simulate the entire growth process involving binary homogeneous nucleation, binary heterogeneous co-condensation, and coagulation among nanoparticles with different compositions, numerical experiments were carried out by changing only the saturation pressure of one material artificially to highlight the effect of the saturation pressure difference between a metal and silicon on the nanopowder formation in thermal plasma fabrication. A greater difference in saturation pressures causes a larger time lag of co-condensation of two material vapors during the collective growth of the metal–silicide nanopowder. The longer time lag of co-condensation results in a wider range of compositions for the mature nanopowder.

Acknowledgments: This work was partly supported by the Japan Society for the Promotion of Science Grant-in-Aid for Scientific Research (B) (KAKENHI: Grant No. 15H03919).

Author Contributions: Masaya Shigeta and Takayuki Watanabe conceived and designed the computational experiment. Masaya Shigeta developed the model and algorithm and performed the computation. Masaya Shigeta and Takayuki Watanabe analyzed the data. Masaya Shigeta wrote the paper.

References

1. Shigeta, M.; Murphy, A.B. Thermal Plasmas for Nanofabrication. *J. Phys. D Appl. Phys.* **2011**, *44*, 174025. [CrossRef]
2. Siegel, R.W. Synthesis and properties of nanophase materials. *Mater. Sci. Eng.* **1993**, *A168*, 189–197. [CrossRef]
3. Matsuura, K.; Hasegawa, T.; Ohmi, T.; Kudoh, M. Synthesis of $MoSi_2$-$TiSi_2$ Pseudobinary Alloys by Reactive Sintering. *Metall. Mater. Transact. A* **2000**, *31*, 747–753. [CrossRef]
4. Kambara, M.; Kitayama, A.; Homma, K.; Hideshima, T.; Kaga, M.; Sheem, K.-Y.; Ishida, S.; Yoshida, T. Nano-composite Si particle formation by plasma spraying for negative electrode of Li ion batteries. *J. Appl. Phys.* **2014**, *115*, 143302. [CrossRef]
5. Sato, T.; Shigeta, M.; Kato, D.; Nishiyama, H. Mixing and magnetic effects on a nonequilibrium argon plasma jet. *Int. J. Therm. Sci.* **2001**, *40*, 273–278. [CrossRef]
6. Shigeta, M.; Sato, T.; Nishiyama, H. Computational simulation of a particle-laden RF inductively coupled plasma with seeded potassium. *Int. J. Heat Mass Transf.* **2004**, *47*, 707–716. [CrossRef]
7. Shigeta, M.; Nishiyama, H. Numerical Analysis of Metallic Nanoparticle Synthesis Using RF Inductively Coupled Plasma Flows. *J. Heat Transf.* **2005**, *127*, 1222–1230. [CrossRef]
8. Fan, X.; Ishigaki, T. Critical free energy for nucleation from the congruent melt of $MoSi_2$. *J. Cryst. Growth* **1997**, *171*, 166–173. [CrossRef]
9. Fan, X.; Ishigaki, T.; Sato, Y. Phase formation in molybdenum disilicide powders during in-flight induction plasma treatment. *J. Mater. Res.* **1997**, *12*, 1315–1326. [CrossRef]
10. Watanabe, T.; Itoh, H.; Ishii, Y. Preparation of ultrafine particles of silicon base intermetallic compound by arc plasma method. *Thin Solid Films* **2001**, *390*, 44–50. [CrossRef]
11. Watanabe, T.; Okumiya, H. Formation mechanism of silicide nanoparticles by induction thermal plasmas. *Sci. Technol. Adv. Mater.* **2004**, *5*, 639–646. [CrossRef]
12. Gerile, N.; Kaga, M.; Kambara, M. Synthesis and Characterization of the Plasma Sprayed Si–Ni Composite Powders as Negative Electrode of lithium-ion Batteries. In Proceedings of the 12th Asia Pacific Physics Conference (APPC12), Makuhari, Japan, 14–19 July 2013.
13. Lümmen, N.; Kraska, T. Homogeneous nucleation and growth in iron-platinum vapor investigated by molecular dynamics simulation. *Eur. Phys. J. D* **2007**, *40*, 247–260. [CrossRef]
14. Girshick, S.L.; Chiu, C.-P.; Muno, R.; Wu, C.Y.; Yang, L.; Singh, S.K.; McMurry, P.H. Thermal plasma synthesis of ultrafine iron particles. *J. Aerosol Sci.* **1993**, *24*, 367–382. [CrossRef]
15. Bilodeau, J.F.; Proulx, P. A mathematical model for ultrafine iron powder growth in thermal plasma. *Aerosol Sci. Technol.* **1996**, *24*, 175–189. [CrossRef]
16. Desilets, M.; Bilodeau, J.F.; Proulx, P. Modelling of the reactive synthesis of ultra-fine powders in a thermal plasma reactor. *J. Phys. D Appl. Phys.* **1997**, *30*, 1951–1960. [CrossRef]
17. Cruz, A.C.D.; Munz, R.J. Vapor Phase Synthesis of Fine Particles. *IEEE Transact. Plasma Sci.* **1997**, *25*, 1008–1016. [CrossRef]
18. Murphy, A.B. Formation of titanium nanoparticles from a titanium tetrachloride plasma. *J. Phys. D Appl. Phys.* **2004**, *37*, 2841–2847. [CrossRef]
19. Shigeta, M.; Watanabe, T. Two-dimensional analysis of nanoparticle formation in induction thermal plasmas with counterflow cooling. *Thin Solid Films* **2008**, *516*, 4415–4422. [CrossRef]
20. Shigeta, M.; Watanabe, T. Numerical investigation of cooling effect on platinum nanoparticle formation in inductively coupled thermal plasmas. *J. Appl. Phys.* **2008**, *103*, 074903. [CrossRef]
21. Colombo, V.; Ghedini, E.; Gherardi, M.; Sanibondi, P.; Shigeta, M. A Two-Dimensional nodal model with turbulent effects for the synthesis of Si nano-particles by inductively coupled thermal plasmas. *Plasma Sources Sci. Technol.* **2012**, *21*, 025001. [CrossRef]
22. Shigeta, M.; Watanabe, T. Growth mechanism of silicon-based functional nanoparticles fabricated by inductively coupled thermal plasmas. *J. Phys. D Appl. Phys.* **2007**, *40*, 2407–2419. [CrossRef]
23. Vorobev, A.; Zikanov, O.; Mohanty, P. Modelling of the in-flight synthesis of TaC nanoparticles from liquid precursor in thermal plasma jet. *J. Phys. D Appl. Phys.* **2008**, *41*, 085302. [CrossRef]
24. Vorobev, A.; Zikanov, O.; Mohanty, P. A Co-Condensation Model for In-Flight Synthesis of Metal-Carbide Nanoparticles in Thermal Plasma Jet. *J. Therm. Spray Technol.* **2008**, *17*, 956–965. [CrossRef]

25. Shigeta, M.; Watanabe, T. Two-Directional Nodal Model for Co-Condensation Growth of Multi-Component Nanoparticles in Thermal Plasma Processing. *J. Therm. Spray Technol.* **2009**, *18*, 1022–1037. [CrossRef]
26. Shigeta, M.; Watanabe, T. Growth model of binary alloy nanopowders for thermal plasma synthesis. *J. Appl. Phys.* **2010**, *108*, 043306. [CrossRef]
27. Cheng, Y.; Shigeta, M.; Choi, S.; Watanabe, T. Formation Mechanism of Titanium Boride Nanoparticles by RF Induction Thermal Plasma. *Chem. Eng. J.* **2012**, *183*, 483–491. [CrossRef]
28. Shigeta, M.; Watanabe, T. Effect of precursor fraction on silicide nanopowder growth under thermal plasma conditions: a computational study. *Powder Technol.* **2016**, *288*, 191–201. [CrossRef]
29. Shigeta, M.; Watanabe, T. Numerical analysis for co-condensation processes in silicide nanoparticle synthesis using induction thermal plasma at atmospheric pressure conditions. *J. Mater. Res.* **2005**, *20*, 2801–2811. [CrossRef]
30. The Japan Institute of Metals. *Metal Data Book*; Maruzen: Tokyo, Japan, 1993; pp. 10–91. (In Japanese)
31. Wyslouzil, B.E.; Wilemski, G. Binary nucleation kinetics. II. Numerical solution of the birth–death equations. *J. Chem. Phys.* **1995**, *103*, 1137–1151. [CrossRef]
32. Massalski, T.B. *Binary Alloy Phase Diagrams*, 2nd ed.; American Society for Metals: Materials Park, OH, USA, 1990; pp. 3367–3371.
33. Wautelet, M.; Dauchot, J.P.; Hecq, M. Phase diagrams of small particles of binary systems: A theoretical approach. *Nanotechnology* **2000**, *11*, 6–9. [CrossRef]
34. Friedlander, S.K. *Smoke, Dust and Haze, Fundamentals of Aerosol Dynamics*, 2nd ed.; Oxford University Press: New York, NY, USA, 2000; p. 280.

![nanomaterials logo] *nanomaterials*

MDPI

Article

Dielectric Barrier Discharge (DBD) Plasma Assisted Synthesis of Ag$_2$O Nanomaterials and Ag$_2$O/RuO$_2$ Nanocomposites

Antony Ananth and Young Sun Mok *

Plasma Applications Laboratory, Department of Chemical and Biological Engineering, Jeju National University, Jeju 690-756, Korea; sebastiananth@gmail.com
* Correspondence: smokie@jejunu.ac.kr; Tel.: +82-64-754-3682; Fax: +82-64-755-3670

Academic Editors: Krasimir Vasilev, Melanie Ramiasa and Thomas Nann
Received: 24 December 2015; Accepted: 22 February 2016; Published: 26 February 2016

Abstract: Silver oxide, ruthenium oxide nanomaterials and its composites are widely used in a variety of applications. Plasma-mediated synthesis is one of the emerging technologies to prepare nanomaterials with desired physicochemical properties. In this study, dielectric barrier discharge (DBD) plasma was used to synthesize Ag$_2$O and Ag$_2$O/RuO$_2$ nanocomposite materials. The prepared materials showed good crystallinity. The surface morphology of the Ag$_2$O exhibited "garland-like" features, and it changed to "flower-like" and "leaf-like" at different NaOH concentrations. The Ag$_2$O/RuO$_2$ composite showed mixed structures of aggregated Ag$_2$O and sheet-like RuO$_2$. Mechanisms governing the material's growth under atmospheric pressure plasma were proposed. Chemical analysis was performed using Fourier transform infrared spectroscopy (FTIR) and X-ray photoelectron spectroscopy (XPS). Thermogravimetric analysis (TGA) showed the thermal decomposition behavior and the oxygen release pattern.

Keywords: atmospheric pressure plasma; DBD; silver oxide; ruthenium oxide; nanomaterials; nanocomposite

---

1. Introduction

Silver oxide nanomaterials (Ag$_2$O NMs) are widely used in catalysis [1,2], sensors [3], preparation of antimicrobial materials [4], drinking water-related applications [5], *etc.* It is an intrinsic p-type semiconductor with a band gap of around 1.5 eV. On account of its high catalytic activity under mild reaction condition, usage of Ag$_2$O NMs is rapidly increasing in many chemical reactions. The efficiency of a material in an application can further be improved when it is mixed with other materials. Such carefully selected nanocomposite materials (two or more metallic or metal oxide structures having the dimension of nanoscale) having excellent individual properties exhibit new and unique characteristics without affecting each other. For example, Lee *et al.* [6] have documented that Ag$_2$O/RuO$_2$ composite exhibits higher capacitance as compared to its individual performances. Ruthenium (IV) oxide (RuO$_2$) itself is an excellent candidate material for catalysis, field emission displays, fuel cells and supercapacitor applications [6–8]. There are varieties of methods such as wet chemical [9], biological [10], thermal deposition [11], thin film-based and recently plasma-mediated routes [12–17] to prepare nanomaterials are currently used. The main advantage of applying plasma-based techniques to synthesize nanomaterials is the possibility of controlling the growth, surface chemistry and surface morphology. For instance, plasma-mediated synthesis does not require stabilizer molecules to prevent the aggregation (steric stabilization) which is commonly encountered in wet chemical synthesis. This is because the growth of material and the control of aggregation are entirely driven by the physical properties such as electrostatic repulsion, electric potential and conductivity (both electrical and thermal) [18]. The surface chemistry (metallic and oxide formations) is mostly decided by the composition of the feed gas used for the generation of plasma. Depending on the gas composition, the density of the active species such as electrons, ions and radicals vary and its interaction with the nucleating particles decides the chemical nature of the surface. The drawbacks of plasma techniques may be electrical safety of high voltage systems and difficulty in mass production.

Unlike other vacuum-based (for example, chemical vapor deposition (CVD), plasma enhanced CVD, *etc.*) or high-temperature plasma techniques, the dielectric barrier discharge (DBD) plasma-based method is non-thermal in nature, works at atmospheric pressure, and can operate at relatively low energy input, which facilitates the production of nanomaterials with unique structural features [8]. In addition, the construction of DBD plasma reactor is relatively easier than any other plasma techniques and the DBD plasma gives reproducible results. In our previous reports [8,14], RuO_2 NMs were synthesized in a DBD plasma reactor, and it was found that the gas composition and the substrate materials used for the growth of NMs played main roles in determining the morphology of the NMs. To our knowledge, the investigation on the growth of composite materials having two different physicochemical properties in the presence of atmospheric pressure plasma is scarce in the literature [18,19]. With this background, this work deals with the syntheses of Ag_2O NMs and Ag_2O/RuO_2 nanocomposite materials in the presence of DBD plasma. Investigation on the surface morphological control, growth mechanism and surface chemical analyses of the plasma-assisted materials would help to understand the importance of this method. The Ag_2O/RuO_2 composite material was chosen based on its potential environmental applications.

2. Results and Discussion

2.1. X-Ray Diffraction Study for Structural Analysis

The X-ray diffraction (XRD) spectrum of Ag_2O NMs is given in Figure 1a. The peaks observed at the diffraction angle (2θ) 26.5°, 32.75°, 37.9°, 54.8°, 65.3° and 68.7° correspond to (110), (111), (200), (220), (311) and (222) set of lattice planes (cubic structure), respectively [JCPDS card No. 76-1393]. The spectrum did not contain any peaks corresponding to impurities or metallic forms, thus showing the Ag_2O NMs of high purity [20]. The high intense peak (111) may refer to the arrangement of lattice atoms in an ordered structural fashion. The XRD spectra in Figure 1b shows the pattern corresponding to Ag_2O/RuO_2 nanocomposite, in which the Miller indices representing Ag_2O and RuO_2 are marked with (#) and (*). Slight shift in the diffraction peak (200), (220) and (311) of Ag_2O was noted at the 2θ angle 38.21°, 54.26° and 64.5°, respectively. This kind of shift (decrease or increase) in the diffraction angle results from the difference between Ag_2O and RuO_2 lattice constants and lattice strain [21]. Moreover, the intensity of (111) peak decreased to a great extent and (200) increased, which may indicate a structural change (especially it may be due to the morphology, since the XRD peak intensity is closely connected with crystal morphology [22]). In comparison, the peaks corresponding to rutile type RuO_2 were clearly observed without any shift [9].

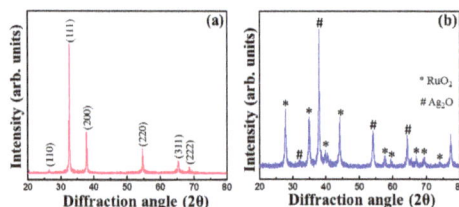

Figure 1. X-ray diffractograms of (**a**) Ag_2O nanomaterials (NMs); and (**b**) Ag_2O/RuO_2 nanocomposite. Arb. units stand for arbitrary units.

2.2. Surface Morphology and Elemental Analysis

Figure 2a,b show the field emission scanning electron microscope (FESEM) surface morphological images of the plasma-synthesized Ag_2O and Ag_2O/RuO_2 nanocomposite powders, respectively. The Ag_2O NMs exhibited bundles of spherical nanoparticles (with particle diameter < 50 nm) whereas the Ag_2O/RuO_2 composite showed the mixed structures of RuO_2 nanosheet and aggregated Ag_2O. The high resolution transmission electron microscope (TEM) images also confirmed the above observation (Figure 2c,d). The elemental composition (atomic percentage) measured by energy dispersive X-ray spectroscopy (EDX) (Figure 2e) for Ag_2O consisted mainly of Ag (96.3%) and O (3.7%). The peaks corresponding to Ag_2O/RuO_2 nanocomposite consisted of Ag (19.8%), Ru (53.7%) and O (26.5%) (Figure 2f). The other peaks found in the spectrum refer to copper and carbon of the TEM grid. For information, the EDX analysis gives the information on the elements present in the given sample, but the exact content of each element in the sample cannot be determined by this technique because the elemental composition depends largely on the scan area.

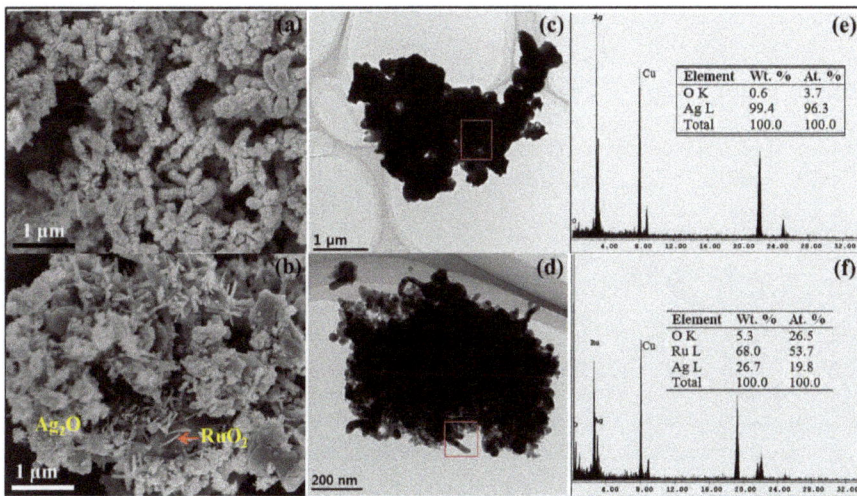

Figure 2. Field emission scanning electron microscope (FESEM) surface morphological images of (a) Ag_2O nanomaterials (NMs); and (b) Ag_2O/RuO_2; high resolution transmission electron microscope (TEM) images of (c) Ag_2O NMs; and (d) Ag_2O/RuO_2; and energy dispersive X-ray spectroscopy (EDX) spectra with the elemental composition for (e) Ag_2O NMs; and (f) Ag_2O/RuO_2.

2.3. Particle Size Analysis

From particle size analysis, size distribution of nanomaterials in an aqueous media can be obtained. This property is of particular importance, considering the applications such as photocatalysis, anti-microbial studies, *etc.* Particles of different size exert different Brownian motion in solution, and thus, when the particles are illuminated by a beam of light, the scattered light fluctuates in response to the individual particles. This fluctuation is monitored and using photon detection method, the particle size distribution is obtained. Particle size analyses were performed for Ag_2O and Ag_2O/RuO_2 nanocomposite, and the results are shown in Figure 3 where the size (nm) of the nanomaterials *versus* light scattering distribution (probability) along with the cumulative distribution is represented. The cumulative distribution refers to the probability that the particle size is less than or equal to a particular size, *i.e.*, accumulated probability up to a particular size. The average hydrodynamic diameters (Stokes diameter) of Ag_2O and Ag_2O/RuO_2 were found to be 44.2 and 485 nm, respectively. Since the Ag_2O NMs showed uniform growth and morphology, more number of particles was observed within 100 nm range. The particles with larger size, especially in Ag_2O/RuO_2 composite may reflect the RuO_2 length. Besides, the Ag_2O NMs did not grow uniformly and exhibited more aggregation in the composite, resulting in bigger size distribution. Average size of the Ag_2O and Ag_2O/RuO_2 nanocomposite obtained from the particle size analysis agreed well with the morphological features obtained by FESEM.

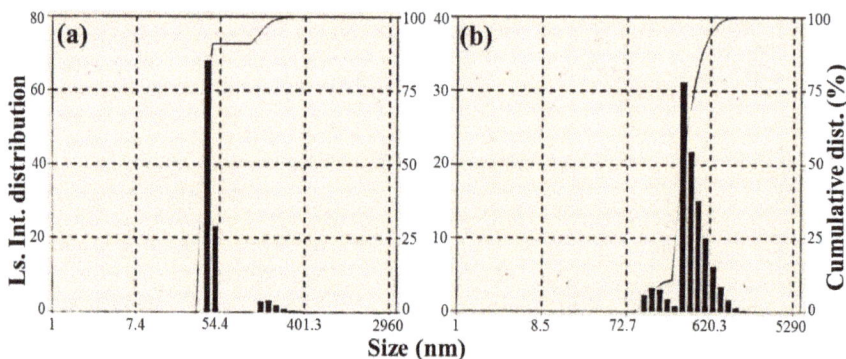

Figure 3. Particle size distributions of (**a**) Ag_2O NMs (b); and (**b**) Ag_2O/RuO_2 nanocomposite. Ls. Int. distribution stands for light scattering intensity distribution.

2.4. Effect of NaOH Concentration on the Morphology and Growth Mechanism

In the previous study [18], it has been reported that materials having higher electrical and thermal conductivities (for example, CuO) usually produce spherical particles in the presence of DBD plasma. Initially, supersaturation of the solution is attained as a result of the evaporation of water due to the heat generated by plasma. Then the accumulation of charged species on the surface of the growing particles favors the growth towards the oppositely charged electrode. Since the incoming electrons are highly mobile on the entire surface of the nucleated Ag_2O species (Ag_2O and CuO band gap values are almost similar, and thus the Ag_2O follows similar growth pattern to CuO NMs), the alternating voltage does not produce any structural growth specifically. This is schematically proposed in Figure 4a (a1). In such cases, changing the NaOH concentration (pH increase) would be an alternative rather than altering the plasma parameters. Once the supersaturation is attained, the growth speed (slow or fast) of the nucleated particles may depend on the NaOH concentration. If the NaOH concentration is increased, attainment of zero charge molecules and formation of hydrated hydroxides starts earlier due to more OH^- ions from NaOH, which may result in bigger particles of different morphologies (Figure 4a (a2)). The mechanism was found suitable by observing the FESEM surface morphology of the Ag_2O NMs obtained with 0.75 and 1.0 M NaOH concentrations, respectively (Figure 5a,b). When the concentration of NaOH was increased from 0.75 to 1.0 M, the length and thickness of the particles also increased, resulting in a flower-like and leaf-like morphology, respectively. On the other hand, the synthesis of RuO_2 NMs (band gap of about 2.5 eV) by DBD plasma resulted in one dimensional structure such as nanosheets (see Figure S1b in Supplementary Materials) due to the potential gradient inside the reactor [15]. Accumulation of static electric charges results in the stacking and compression of the nucleated particles in response to the created electric potential, leading to sheet-like structures (Figure 4b). On the other hand, the length of the RuO_2 nanosheet and the aggregation behavior of Ag_2O were influenced to a great extent in the Ag_2O/RuO_2 nanocomposite when increasing the NaOH concentration (Figures 4c and 5c,d). The nanocomposite exhibits mixed structures due to the competition between the nucleation and growth of the starting materials (one particle growth may suppress or hinder the other). It is learnt that the uniform growth with specific morphological features could be obtained if the NMs are prepared separately.

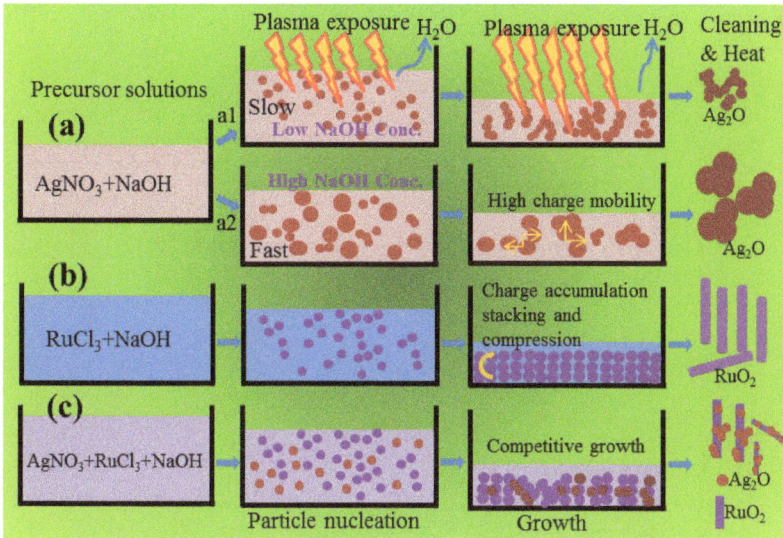

Figure 4. The proposed growth mechanisms for (**a**) Ag$_2$O and (**b**) RuO$_2$ nanomaterials and (**c**) Ag$_2$O/RuO$_2$ nanocomposite in the presence of plasma. The **a1** and **a2** represent the Ag$_2$O growth at 0.5 and 1.0 M NaOH concentration.

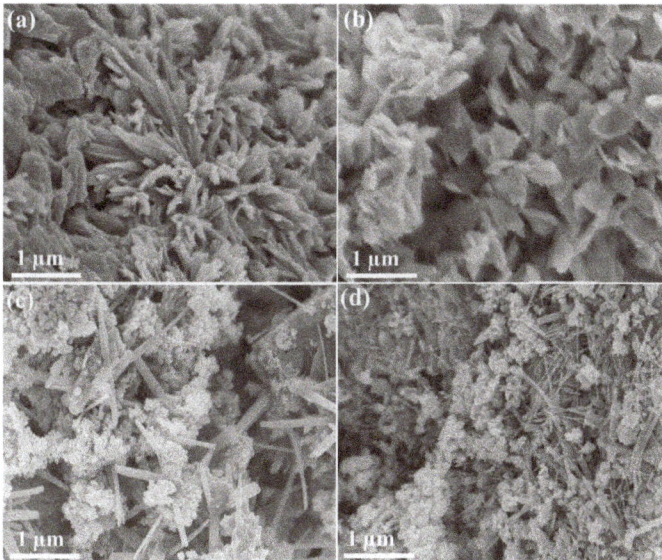

Figure 5. Morphology of Ag$_2$O NMs obtained with an NaOH concentration of (**a**) 0.75 M; and (**b**) 1.0 M; and morphology of Ag$_2$O/RuO$_2$ nanocomposite obtained with an NaOH concentration of (**c**) 0.75 M; and (**d**) 1.0 M.

2.5. Bulk and Surface Chemical Analysis

Figure 6 shows the Fourier transform infrared (FTIR) spectra of the Ag$_2$O NMs. The Ag$_2$O/RuO$_2$ nanocomposite exhibited almost similar spectrum and thus not shown here. The functional groups corresponding to Ag_2O nanoparticles (Ag–O bonds) were observed around 550 and 720 cm^{-1} (highlighted) referring to its transverse and longitudinal optical phonon vibrational frequencies, respectively [11,23,24]. The signatures at 1638 and 3400 cm^{-1} correspond to the vibration of OH. The metal-oxygen vibrations for Ag–O and Ru–O in Ag_2O/RuO_2 nanocomposite were observed between 650 and 800 cm^{-1}. In this case, the identification of the peaks corresponding to Ag–O and Ru–O seems difficult since the composite exhibited identical FTIR spectra as observed for Ag_2O.

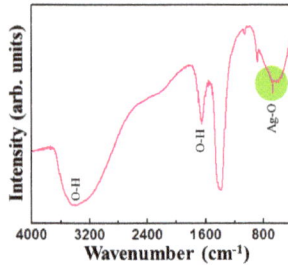

Figure 6. Fourier transform infrared spectroscopy (FTIR) spectrum of Ag_2O NMs.

In order to confirm the presence of chemical functional groups in a detailed manner, the X-ray photoelectron spectra (XPS) of Ag_2O NMs and Ag_2O/RuO_2 nanocomposite were taken and given in Figure 7a,b. The Ag_2O spectra showed the presence of Ag (as 3d and 3p orbital splitting), oxygen and adventitious carbon signals [25,26]. The Ag_2O/RuO_2 nanocomposite exhibited the presence of ruthenium, silver and oxygen. The high resolution core level XPS spectra of O 1s, Ag 3d, C 1s and Ru 3p spectra corresponding to the samples Ag_2O and Ag_2O/RuO_2 are given in Figure 8. The O 1s spectrum of Ag_2O contains two species at the binding energy 531.5 and 529.8 eV referring to oxygen from adsorbed water and Ag_2O [27–31], which are noted with its composition (atomic %). The O 1s spectrum of Ag_2O/RuO_2 contains one more peak corresponding to RuO_x in addition to the above (Figure 8b) [32]. The presence of other elements were found out by Gauss-Laurentian peak fitting program and listed in Table 1. The Ag spectra of the Ag_2O and Ag_2O/RuO_2 nanocomposite consisted of a doublet peak at the binding energy 368.2 eV ($3d_{5/2}$) and 374 eV ($3d_{3/2}$) with the doublet separation of 5.8 eV (Figure 7c,d). The Ru 3p spectrum of Ag_2O/RuO_2 contains two regions corresponding to RuO_2 and higher oxides of ruthenium (RuO_x, where $x = 3$, probably) [33].

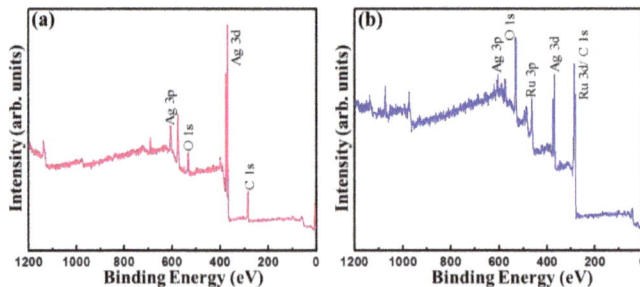

Figure 7. X-ray photoelectron spectroscopy (XPS) survey spectra of (a) Ag_2O NMs (b); and (b) Ag_2O/RuO_2 nanocomposite.

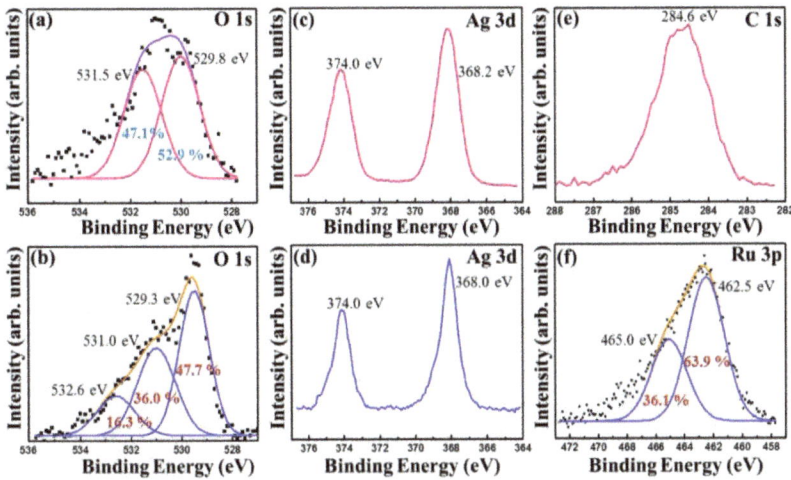

Figure 8. High-resolution core level XPS spectra of (a) O 1s in Ag$_2$O; (b) O 1s in Ag$_2$O/RuO$_2$ nanocomposite; (c) Ag 3d in Ag$_2$O; (d) Ag 3d in Ag$_2$O/RuO$_2$ nanocomposite; (e) C 1s in Ag$_2$O; and (f) Ru 3p in Ag$_2$O/RuO$_2$ nanocomposite.

Table 1. Results of the X-ray photoelectron spectroscopy (XPS) analysis.

Binding Energy (eV)	Species Assignment	Reference
284.6	Carbon	[25]
368.2, 374.0, 529.3, 529.8	Ag$_2$O	[27–31]
462.5	RuO$_2$	[32]
531.5	Oxygen from adsorbed water	[31]
531.0, 465.0	RuO$_x$/Ru	[33]
532.6	Diffused oxygen atoms	[26]

2.6. Thermal Study

The thermo-gravimetric analysis (TGA) technique is fundamentally used to track the changes in physical (phase transformation, crystallization) or chemical (redox reactions) process that occurs when heating a material. The decomposition reaction of Ag$_2$O and Ag$_2$O/RuO$_2$ in the presence of N$_2$ atmosphere can be expressed by the following equations. It is assumed that the sample contains stoichiometric oxygen and the weight loss is proportional to the amount of oxygen present originally.

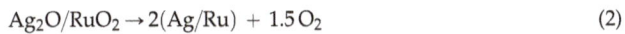

$$Ag_2O \rightarrow 2Ag + 0.5 O_2 \qquad (1)$$

$$Ag_2O/RuO_2 \rightarrow 2(Ag/Ru) + 1.5 O_2 \qquad (2)$$

The oxygen loss associated with the decomposition is as follows:

$$\Delta W = \frac{\text{Atomic or molecular weight of oxygen}}{\text{Molecular weight of metal oxide (s)}} \text{Actual weight of metal oxide (s)} \qquad (3)$$

Figure 9a shows TGA thermogram of Ag$_2$O NMs. It showed two stage decomposition or transformation starting at 194.14 °C and 386.82 °C, which could account for the decomposition of water (and also AgO) and the conversion from Ag$_2$O to metallic Ag [34] with the activation energy of 283.4 kJ·mol^{-1} at the standard atmosphere [35]. Since the decomposition was rapid, about 10.25 wt % of the sample was lost at the end of the reaction. According to the above equations, the sample contains initially 6.89 wt % oxygen and final loss amounts to 6.19 wt %. This loss was observed up to 505.68 °C and beyond it the sample exhibited molten salt or ionic liquid [36] behavior with small percentage of oxygen (0.7%, see Table 2). But the Ag_2O/RuO_2 nanocomposite (Figure 9b) showed multi-stage decomposition phenomena. Initial water removal was observed until the temperature reached at 150 °C, followed by maximum Ag_2O decomposition at around 382.26 °C and eventually conversion into metallic Ru formation above 700 °C with mass loss of 2.1 wt %. On considering the oxygen loss, it exhibited a gradual decomposition with its initial concentration of 13.149 wt % to the final loss 12.873 wt %. The Ag_2O/RuO_2 nanocomposite showed a steady oxygen release and maximum decomposition to metallic nature.

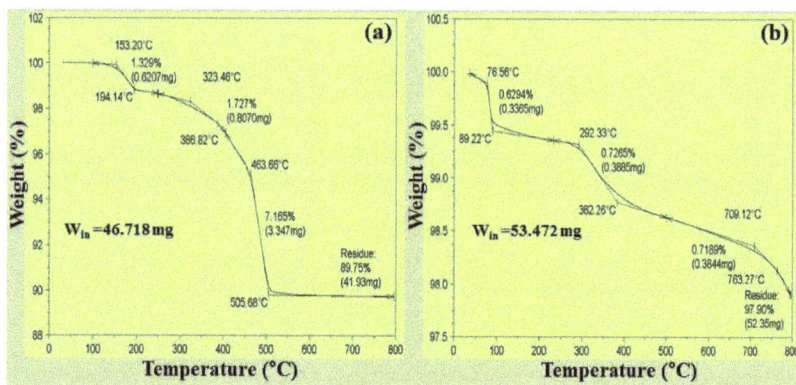

Figure 9. Thermogravimetric analysis (TGA) thermogram of (a) Ag_2O NMs and (b) Ag_2O/RuO_2 nanocomposite (b). W_{in} : initial weight of the sample.

Table 2. Amount of oxygen before and after Thermogravimetric analysis (TGA) analysis.

Sample	Initial Sample Mass (mg)	Initial O_2 Mass (mg)	Final Sample Mass (mg)	Lost O_2 Mass (mg)	Remaining O_2 Mass (mg)	Remaining O_2 Mass (%)
Ag_2O	46.718	3.22	41.93	2.89	0.330	0.700
Ag_2O/RuO_2	53.472	7.031	52.35	6.883	0.148	0.276
* RuO_2	42.988	10.388	40.84	9.822	0.566	1.192

* Supplementary Materials.

3. Experimental Section

3.1. DBD Plasma Reactor Setup

The NMs were prepared in a self-designed, rectangular box type DBD plasma reactor [8,15], the photographic image of which is given in Figure S2. In brief, two electrodes (4 cm wide and 15 cm long) acted as high voltage and ground electrodes which were covered with glass dielectric of 1.5 mm thickness. The DBD plasma reactor prepared as above was fixed in an acrylic chamber with gas inlet and outlet. The discharge gap (distance between high voltage and ground electrodes) was fixed at 1.5 cm and the nanomaterial precursor solution was kept inside. Alternating current (AC) high voltage (operating frequency: 400 Hz) was applied across the electrodes to generate the plasma. Since argon was used, plasma consisting of numerous filamentary microdischarges was readily created despite the large discharge gap.

3.2. Material Synthesis

Precursor chemicals such as AgNO$_3$ (MW 169.87 g mol^{-1}, Sigma-Aldrich, St. Louis, MO, USA), RuCl$_3 \cdot x$H$_2$O (Sigma-Aldrich, St. Louis, MO, USA) and NaOH (Shinyo Pure Chemical Co., Osaka, Japan) were used as received. Exactly 0.5 M AgNO$_3$, RuCl$_3 \cdot x$H$_2$O and NaOH solutions were prepared separately. For the preparation of Ag$_2$O NMs, 500 µL of the precursor solution was mixed with 500 µL NaOH, agitated for 2 min, spread on the glass substrate and kept inside the DBD plasma reactor [14]. For Ag$_2$O/RuO$_2$ nanocomposite, 500 µL respective precursor solutions and 1 mL NaOH were used. Gas purging with Ar gas was done for 15 min and the solution was exposed to plasma. The input power was maintained at 38.20 W for all the samples. The input power was measured by a digital power meter (Model WT200, Yokogawa, Tokyo, Japan). Depending on the sample dryness inside the plasma region, the discharge voltage fluctuated in the range of 16–20 kV, despite the same input power. The temperature inside the reactor was found to be varying from room temperature at starting to 70–76 °C at the end of the reaction. According to our previous study, conventional wet chemical synthesis produces aggregated spherical type of NMs [9]. At the end of 3-h plasma reaction, the solid powder was collected and washed repeatedly with deionized water. Heat treatment was performed in a furnace at 200 °C for 24 h.

3.3. Material Characterization

Crystalline nature of the materials was studied using an X-ray diffractometer (D/Max Ultima III, Rigaku Corp., Tokyo, Japan) fitted with a monochromatic Cu K$_\alpha$ radiation (wavelength, λ = 0.154 nm) operated at 40 mA and 40 kV. The surface morphology was observed by an FESEM (JEM 1200 EX II, JEOL, Tokyo, Japan) and TEM. The elemental composition and the nanomaterial's size were analyzed using an EDX (Model: R-TEM, CM200-UT, Philips, Ventura, CA, USA) and a particle size analyzer (ELS8000, Otsuka Electronics, Osaka, Japan), respectively. FTIR spectroscopy (IFS 66/S, Bruker, Bremen, Germany) was used to study the chemical functional groups. Chemical nature of the surface was also characterized by using an XPS (ESCA 2000, VG Microtech, East Grinstead, UK) with monochromatic Mg K$_\alpha$ X-ray radiation (1253.6 eV) operated with a 13 kV and 15 mA excitation source. TGA was carried out to study the thermal decomposition behavior of the prepared NMs. The samples were heated at a ramping rate of 10 °C· min^{-1} and simultaneous weight loss was recorded precisely (Q50 Analyzer, version 20.10, TA instruments, New Castle, DE, USA).

4. Conclusions

The DBD plasma-mediated syntheses of Ag$_2$O NMs and Ag$_2$O/RuO$_2$ nanocomposite were investigated. Both materials exhibited crystalline nature with a slight shift in the diffraction angle for Ag$_2$O in the Ag$_2$O/RuO$_2$ nanocomposite. The Ag$_2$O exhibited unique morphology of spherical bundles whereas the composite showed the mixture of RuO$_2$ nanorod and Ag$_2$O spherical aggregates. The Ag$_2$O morphology was significantly altered by changing the concentration of NaOH. The surface chemical analyses by XPS revealed that the sample consisted mainly of Ag$_2$O. The ruthenium in the Ag$_2$O/RuO$_2$ nanocomposite was in two oxide forms such as RuO$_2$ and RuO$_x$. The thermal study showed a fast transformation of Ag$_2$O into metallic Ag in the temperature range of 200–500 °C. But the Ag$_2$O/RuO$_2$ nanocomposite exhibited gradual decomposition without major change in the initial weight until 700 °C. This nanocomposite may exhibit superior redox properties even at high temperature and thus it would serve as a potential candidate in many applications.

Supplementary Materials: The following are available online at http://www.mdpi.com/2079-4991/6/3/42/s1.

Acknowledgments: This research was supported by Basic Science Research Program through the National Research Foundation of Korea (NRF) funded by the Ministry of Science, ICT and future Planning (Grant No. 2013R1A2A2A01067961).

Author Contributions: Antony Ananth performed the experimental work and analyzed the data; Young Sun Mok supervised all the study and participated in the interpretation of the results.

Conflicts of Interest: The authors declare no conflict of interest.

References

1. Wang, X.; Li, S.; Yu, H.; Yu, J.; Liu, S. Ag$_2$O as a new visible-light photocatalysts: Self-stability and high photocatalytic activity. *Chem. Eur. J.* **2011**, *17*, 7777–7780. [CrossRef] [PubMed]
2. Wang, W.; Zhao, Q.; Dong, J.; Li, J. A novel silver oxides oxygen evolving catalyst for water splitting. *Int. J. Hydrogen Energy* **2011**, *36*, 7374–7380. [CrossRef]
3. Rahman, M.A.; Khan, S.B.; Jamal, A.; Faisal, M.; Asiri, A.M. Fabrication of highly sensitive acetone sensor based on sonochemically prepared as-grown Ag$_2$O nanostructures. *Chem. Eng. J.* **2012**, *192*, 122–128. [CrossRef]
4. Jin, Y.; Dai, Z.; Liu, F.; Kim, H.; Tong, M.; Hou, Y. Bactericidal mechanisms of Ag$_2$O/TNBs under both dark and light conditions. *Water Res.* **2013**, *47*, 1837–1847. [CrossRef] [PubMed]
5. Fu, H.-T.; Zhao, L.-M.; Luo, M.; Zhang, H.-Y.; Zhang, J. Determination of Chloracetic Acids in Drinking Water by Ion Chromatography Using Silver Oxide as Precipitant Eliminating Interference of Chlorate in Matrix. *Chin. J. Anal. Chem.* **2008**, *36*, 1407–1410. [CrossRef]
6. Lee, Y.; Ye, B.-U.; Yu, H.K.; Lee, J.-L.; Kim, M.H.; Baik, J.M. Facile Synthesis of Single Crystalline Metallic RuO$_2$ Nanowires and Electromigration-Induced Transport Properties. *J. Phys. Chem. C* **2003**, *115*, 4611–4615. [CrossRef]
7. Lee, K.-Y.; Chen, C.-A.; Lian, H.-B.; Chen, Y.-M.; Huang, Y.-S.; Keisser, G. Pattern, growth and field emission characteristics of flower-like RuO$_2$ nanostructures. *Jpn. J. Appl. Phys.* **2010**, *49*, 105002–105005. [CrossRef]
8. Ananth, A.; Mok, Y.S. Synthesis of RuO$_2$ nanomaterials under dielectric barrier discharge plasma at atmospheric pressure-influence of substrates on the morphology and application. *Chem. Eng. J.* **2014**, *239*, 290–298. [CrossRef]
9. Ananth, A.; Dharaneedharan, S.; Gandhi, M.S.; Heo, M.-S.; Mok, Y.S. Novel RuO$_2$ nanosheets–facile synthesis, characterization and application. *Chem. Eng. J.* **2013**, *223*, 729–736. [CrossRef]
10. Ananth, A.; Arthanareeswaran, G.; Ismail, A.F.; Mok, Y.S.; Matsura, T. Effect of bio-mediated route synthesized silver nanoparticles for modification of polyethersulfone membranes. *Colloids Surf. A* **2014**, *451*, 151–160. [CrossRef]
11. Hosseinpour-Mashkani, S.M.; Ramezani, M. Silver and Silver-Oxide Nanoparticles: Synthesis and Characterization by Thermal Evaporation. *Mater. Lett.* **2014**, *130*, 259–262. [CrossRef]
12. Wei, Y.; Zuo, X.; Li, X.; Song, S.; Chen, L.; Shen, J.; Meng, Y.; Zhao, Y.; Fang, S. Dry Plasma Synthesis of Graphene Oxide–Ag Nanocomposites: A Simple and Green Approach. *Mater. Res. Bull.* **2014**, *53*, 145–150. [CrossRef]
13. Li, Y.; Zhang, Y.; Fu, H.; Wang, Z.; Li, X. Plasma-assisted speedy synthesis of mesoporous Ag$_2$O nanotube. *Mater. Lett.* **2014**, *126*, 131–134. [CrossRef]
14. Kim, D.W.; Park, D.-H. Preparation of indium tin oxide (ITO) nanoparticles by DC arc plasma. *Surf. Coat. Technol.* **2010**, *205*, S201–S205. [CrossRef]
15. Ananth, A.; Gandhi, M.S.; Mok, Y.S. Dielectric barrier discharge (DBD) plasma reactor: An efficient tool to prepare novel RuO$_2$ nanorods. *J. Phys. D* **2013**, *46*, 155202–155209. [CrossRef]
16. Marino, E.; Huijser, T.; Creyghton, Y.; Hejiden, A.V.D. Syntheis and coating of copper oxide nanoparticles using atmospheric pressureplasmas. *Surf. Coat. Technol.* **2007**, *201*, 9205–9208. [CrossRef]
17. Pootawang, P.; Saito, N.; Takai, O.; Lee, S.Y. Rapid synthesis of ordered hexagonal mesoporous silica and their incorporation with Ag nanoparticles by solution plasma. *Mater. Res. Bull.* **2012**, *47*, 2726–2729. [CrossRef]
18. Ananth, A.; Mok, Y.S. Dielectric barrier discharge plasma-mediated synthesis of several oxide nanomaterials and its characterization. *Powder Technol.* **2015**, *269*, 259–266. [CrossRef]
19. Choi, S.; Lee, M.-S.; Park, D.-W. Photocatalytic performance of TiO$_2$/V$_2$O$_5$ nanocomposite powder prepared by DC arc plasma. *Curr. Appl. Phys.* **2014**, *14*, 433–438. [CrossRef]
20. Sullivan, K.T.; Wu, C.; Piekiel, N.W.; Gaskell, K.; Zachariah, M.R. Synthesis and reactivity of nano-Ag$_2$O as an oxidizer for energetic systems yielding antimicrobial products. *Combust. Flame* **2013**, *160*, 438–446. [CrossRef]

21. Gultekin, D.; Alaf, M.; Akbulut, H. Synthesis and characterization of ZnO nanopowders and ZnO-CNT nanocomposites prepared by chemical precipitation route. *Acta Phys. Polonica A* **2013**, *123*, 274–276. [CrossRef]

22. Inoue, M; Hirasawa, I. The relationship between crystal morphology and XRD peak intensity on $CaSO_4 \cdot 2H_2O$. *J. Cryst. Growth* **2013**, *380*, 169–175.

23. Banerjee, S.; Maity, A.K.; Chakravorty, D. Quantum confinement effect in heat treated silver oxide nanoparticles. *J. Appl. Phys.* **2000**, *87*, 8541–8544. [CrossRef]

24. Yong, N.L.; Ahmad, A.; Mohammad, A.W. Synthesis and characterization of silver oxide nanoparticles by a novel method. *Int. J. Sci. Eng. Res.* **2013**, *4*, 155–158.

25. Xie, Y.; Sherwood, P.M.A. Ultrahigh purity graphite electrode by core level and valence band XPS. *Surf. Sci. Spectra* **1992**, *1*, 367–372. [CrossRef]

26. Borman, V.D.; Gusev, E.P.; Lebedinski, Y.Y.; Troyan, V.I. Mechanism of submonolayer oxide formation on silicon surfaces upon thermal oxidation. *Phys. Rev. B* **1994**, *49*, 5415–5423. [CrossRef]

27. Rodriguez, J.A. Metal-Metal Bonding on Surfaces: Electronic and Chemical Properties of Ag on Ru(001). *Surf. Sci.* **1993**, *296*, 149–163. [CrossRef]

28. Tjeng, L.H.; Meinders, M.B.J.; van Elp, J.; Ghijsen, J.; Sawatzky, G.A.; Johnson, R.L. Electronic structure of Ag_2O. *Phys. Rev. B* **1990**, *41*, 3190–3199. [CrossRef]

29. Kaushik, V.K. XPS Core Level Spectra and Auger Parameters for Some Silver Compounds. *J. Electron Spectrosc. Relat. Phenom.* **1991**, *56*, 273–277. [CrossRef]

30. Gerenser, L.J. Photoemission investigation of silver/poly(ethylene terephthalate) interfacial chemistry: The effect of oxygen-plasma treatment. *J. Vac. Sci. Technol. A* **1990**, *8*, 3682–3691. [CrossRef]

31. Pashutski, A.; Hoffman, A.; Folkman, M. Low temperature XPS and AES studies of O_2 adsorption on Al(100). *Surf. Sci.* **1989**, *208*, L91–L97. [CrossRef]

32. Sarma, D.D.; Rao, C.N.R. XPES studies of oxides of second- and third-row transition metals including rare earths. *J. Electron Spectrosc. Relat. Phenom.* **1980**, *20*, 25–45. [CrossRef]

33. Shen, J.Y.; Adnot, A.; Kaliaguine, S. An ESCA study of the interaction of oxygen with the surface of ruthenium. *Appl. Surf. Sci.* **1991**, *51*, 47–60. [CrossRef]

34. Jelic, D.; Penavin-Skundric, J.; Majstorovic, D.; Mentus, S. The thermogravimetric study of silver(I) oxide reduction by hydrogen. *Thermochim. Acta* **2011**, *526*, 252–256. [CrossRef]

35. Waterhouse, G.I.N.; Bowmaker, G.A.; Metson, J.B. The thermal decomposition of silver (I, III) oxide: A combined XRD, FT-IR and Raman spectroscopic study. *Phys. Chem. Chem. Phys.* **2001**, *3*, 3838–3845. [CrossRef]

36. Siddiqui, M.R.H.; Adil, S.F.; Assal, M.E.; Ali, R.; Al-Warthan, A. Synthesis and characterization of silver oxide and silver chloride nanoparticles with high thermal stability. *Asian J. Chem.* **2013**, *25*, 3405–3409.

nanomaterials

MDPI

Article

Developments of the Physical and Electrical Properties of NiCr and NiCrSi Single-Layer and Bi-Layer Nano-Scale Thin-Film Resistors

Huan-Yi Cheng [1], Ying-Chung Chen [1], Chi-Lun Li [1], Pei-Jou Li [1], Mau-Phon Houng [2] and Cheng-Fu Yang [3,*

[1] Department of Electrical Engineering, National Sun Yat-sen University, Kaohsiung 804, Taiwan; eveflora818@gmail.com (H.-Y.C.); ycc@mail.ee.nsysu.edu.tw (Y.-C.C.); insist256@hotmail.com (C.-L.L.); sunnyrain516@hotmail.com (P.-J.L.)
[2] Institute of Microelectronics, National Cheng-Kung University, No.1, University Road, Tainan City 701, Taiwan; mphoung@eembox.ncku.edu.tw
[3] Department of Chemical and Materials Engineering, National University of Kaohsiung, Kaohsiung 81141, Taiwan
* Correspondence: cfyang@nuk.edu.tw; Tel.: +886-7-591-9283; Fax: +886-7-591-9277

Academic Editors: Krasimir Vasilev and Melanie Ramiasa
Received: 9 December 2015; Accepted: 18 February 2016; Published: 25 February 2016

Abstract: In this study, commercial-grade NiCr (80 wt % Ni, 20 wt % Cr) and NiCrSi (55 wt % Ni, 40 wt % Cr, 5 wt % Si) were used as targets and the sputtering method was used to deposit NiCr and NiCrSi thin films on Al_2O_3 and Si substrates at room temperature under different deposition time. X-ray diffraction patterns showed that the NiCr and NiCrSi thin films were amorphous phase, and the field-effect scanning electronic microscope observations showed that only nano-crystalline grains were revealed on the surfaces of the NiCr and NiCrSi thin films. The log (resistivity) values of the NiCr and NiCrSi thin-film resistors decreased approximately linearly as their thicknesses increased. We found that the value of temperature coefficient of resistance (TCR value) of the NiCr thin-film resistors was positive and that of the NiCrSi thin-film resistors was negative. To investigate these thin-film resistors with a low TCR value, we designed a novel bi-layer structure to fabricate the thin-film resistors via two different stacking methods. The bi-layer structures were created by depositing NiCr for 10 min as the upper (or lower) layer and depositing NiCrSi for 10, 30, or 60 min as the lower (or upper) layer. We aim to show that the stacking method had no apparent effect on the resistivity of the NiCr-NiCrSi bi-layer thin-film resistors but had large effect on the TCR value.

Keywords: thin-film resistor; sputtering method; sheet resistance; value of temperature coefficient of resistance (TCR value); Bi-layer

1. Introduction

A wide variety of materials have been investigated as thin-film resistors in integrated circuit (IC) applications. The need for appropriate properties—such as high sheet resistance, low-temperature coefficient of resistance, and stability under ambient conditions—has motivated investigations into electronic conduction mechanisms in a number of ceramal [1,2] and alloy resistor systems [3,4]. In IC fabrication technologies, resistors can be implemented by using diffusion methods fabricated in the base and emitter regions of bipolar transistors, or in the source/drain regions of a CMOS, or by depositing thin films on the surfaces of wafers. Numerous studies have been published on the active metal brazing of engineering ceramics to increase service temperatures [5]. Thompson *et al.* investigated the magnetic properties of a series of NiCr alloys, which possess lower Curie temperatures and lower saturation magnetization. Those alloys have the potential to be developed as suitable alloys for reducing ferromagnetism, and can be biaxially textured [6,7].

NiCr alloys have also for many years been the most common materials as using in metal-based thin-film resistors, so considerable attentions have been focused on understanding the resistivity and temperature coefficient of resistance (TCR) of vacuum-deposited thin films [8,9]. For that, NiCr (80 wt % Ni, 20 wt % Cr) thin films were prepared by radio frequency (RF) magnetron sputter method because the deposition technology was used widely in thin films' preparation, and the NiCr thin films were deposited on silicon (Si) and Al_2O_3 substrates under different deposition time. CrSi-based thin films are a particularly interesting material to be deposited as the thin-film resistors, as they offer certain advantages [10]. For example, they have high sheet resistance, low TCR values, high thermal stability, good long-term reliability, and chemical stability [11]. Dong and colleagues investigated the properties of CrSi (Cr:Si = 1:3) thin-film resistors and found that the resistors doped with 3–6 at % Ni had resistivity 1.31–1.49 times higher than those of CrSi thin films without Ni [12]. Lee and Shin also investigated NiCr-based alloy thin-film resistors with the precision characteristics, using a material comprised of Ni (75 wt %), Cr (20 wt %) Al (3 wt %), Mn (4 wt %), and Si (1 wt %) as target, a DC magnetron as sputtering method, and changing the various deposition parameters, such as power, pressure, substrate temperature, and post-deposition annealing temperature [13].

In the present study, we employed an RF magnetron sputtering method to deposit NiCr (80 wt % Ni, 20 wt % Cr) and NiCrSi (55 wt % Ni, 40 wt % Cr, 5 wt % Si) thin films on Si and Al_2O_3 substrates using different deposition time. We found that even in the amorphous phase, the thicknesses of the NiCr and NiCrSi thin films increased with increasing deposition time. We thoroughly investigated the effects of deposition time on the electrical properties of the NiCr and NiCrSi thin films, and the measured results suggested that deposition time affected their properties, including resistance, resistivity, and TCR. We also found that NiCr thin-film resistors had positive TCR values whereas NiCrSi thin-film resistors had negative TCR values. We therefore examined thin-film resistors with two different bi-layer NiCr and NiCrSi thin-film structures. To create the bi-layers, we fixed the deposition time of the NiCr thin films at 10 min and the deposition time of the NiCrSi thin films at 10, 30, or 60 min. We would show that they could be explored to fabricate the thin-film resistors having TCR values close to 0 ppm/°C. We also showed that the deposition conditions (*i.e.*, whether the NiCr thin films had been deposited as upper or lower layers) and the deposition time (and, hence, the thickness) of the NiCrSi thin films in the bi-layer structures had a large effect on the electrical properties of our NiCr-NiCrSi bi-layer thin-film resistors.

2. Experimental Section

In this study, we used commercial-grade NiCr (80 wt % Ni, 20 wt % Cr) and NiCrSi (55 wt % Ni, 40 wt % Cr, 5 wt % Si) as the targets. We then employed the RF magnetron sputtering method to deposit NiCr and NiCrSi thin films on Al_2O_3 substrates for electrical and thickness measurements and on Si substrates for cross-sectional observation and thickness measurement. The side view of the deposited single-layer NiCr-based and NiCrSi-based thin films is shown in Figure 1a, which is used to measure the physical and electrical properties. A green glass paste on the ceramic, shown in Figure 1b, was used to protect the Al_2O_3 substrate and it was removed after the deposition processes, following ultrasonic cleaning in ethyl alcohol. The length between the two electrodes (also called the length of the thin-film resistors) was 4.0 mm, the width of the thin-film resistors and electrodes was 2.8 mm, and the length of the electrodes was 1.2 mm (see Figure 1a,b). The deposition parameters were: deposition power 150 W, deposition pressure 5 mTorr in a pure Ar atmosphere, and deposition temperature 25 °C. The deposition time was changed from 10 min to 60 min for the NiCr-based thin films and from 10 min to 150 min for the NiCrSi-based thin films. A surface macro image of the fabricated NiCrSi thin-film resistors is shown in Figure 1c. The crystalline structures of the deposited NiCr-based and NiCrSi-based thin films were determined by X-ray diffraction (XRD) (Cu-Kα, Bruker D8, Billerica, MA, USA). The thicknesses of the as-deposited single-layer NiCr and NiCrSi thin films were obtained using α-step equipment and confirmed using field emission scanning electron microscopy (FESEM, Hitachi, Osaka, Japan). FESEM was also employed to determine the surface morphologies of the deposited NiCr-based and NiCrSi-based thin films and to obtain cross-sectional observations and confirm the thicknesses of the bi-layer thin-film structures.

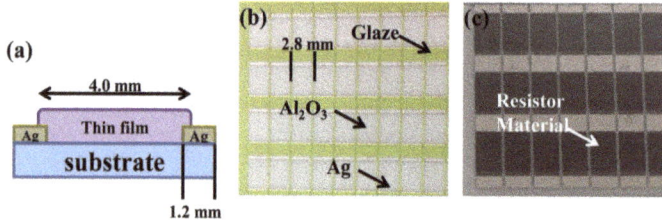

Figure 1. (a) Schematic side view of the thin-film resistors' structure, (b) top view before the protective glass films (green color) were removed, and (c) top view after the protective glass films were removed and the resistor materials (black) were deposited.

The resistances of the NiCr and NiCrSi thin-film resistors were measured using the four-point probe method, and the resistivity was calculated with the measured resistances and thicknesses of the NiCr and NiCrSi thin films, according to Equation (1):

$$R = \rho \times A/l \tag{1}$$

where R is the resistance, ρ is the resistivity, A is the area of the resistor, and l is the length of the resistor.

In this study, the measured temperatures were 25, 50, 75, 100, and 125 °C. The resistivity measured at those temperatures was used to find the TCR values of the NiCr-based and NiCrSi-based thin-film resistors and the two bi-layer thin-film resistors. As already mentioned, we found that the NiCr-based thin-film resistors had positive TCR values whereas the NiCrSi-based thin-film resistors had negative TCR values, so we investigated novel NiCr-NiCrSi bi-layer thin-film resistors, the structures of which are shown in Figure 2. We hoped to develop a thin-film resistor with a TCR value close to 0 ppm/°C, so we created the bi-layer thin-film resistors using two different stacking methods: (i) NiCr thin films were deposited for 10 min as the upper layer, and NiCrSi thin films were deposited for 10, 30, or 60 min as the lower layer (Figure 2a); or (ii) NiCr thin films were deposited for 10 min as the lower layer and NiCrSi thin films were deposited for 10, 30, or 60 min as the upper layer (Figure 2b).

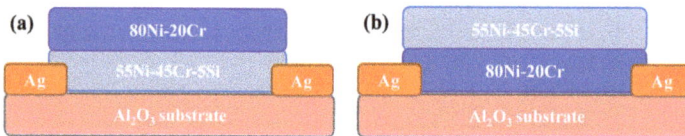

Figure 2. Schematic structures of two different bi-layer NiCr and NiCrSi thin-film resistors. (a) NiCr thin films were deposited for 10 min as the upper layer, and NiCrSi thin films were deposited for 10, 30, or 60 min as the lower layer; (b) NiCr thin films were deposited for 10 min as the lower layer, and NiCrSi thin films were deposited for 10, 30, or 60 min as the upper layer.

To measure each layer's and bi-layer's thicknesses, at least eight samples were deposited at the same time. After deposition, at least three samples were used to measure the thickness of the lower layer, which was also obtained using α-step equipment and confirmed with FESEM. The other five samples were used to deposit the upper layer, and the total thickness of the bi-layer thin films was also measured by the same process. The thickness of the upper layer was calculated using the total thickness minus the thickness of the lower layer and was confirmed by FESEM. The resistance of the bi-layer thin-film resistors was also measured using the four-point probe method, and the resistivity was calculated from the resistance and thickness of the bi-layer structures' thin films. Finally, we measured the TCR of the bi-layer structures' thin-film resistors.

3. Results and Discussion

The effect of deposition time on the thicknesses of the NiCr-based and NiCrSi-based thin films was investigated, and the results are shown in Figure 3. The thicknesses of the NiCr thin films deposited over 10, 30, and 60 min were about 64.3, 170.7, and 327.9 nm, and the thicknesses of the NiCrSi thin films deposited over 10, 30, 60, and 150 min were about 30.8, 90.7, 140.1, and 334.7 nm, respectively. The results in Figure 3 suggest that the deposition rate of the NiCr thin films was higher than that of the NiCrSi thin film. In both instances, however, as the deposition time increased, there was a linear increase in the thicknesses of the NiCr-based and NiCrSi-based thin films.

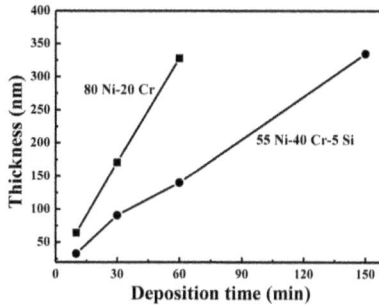

Figure 3. Variations in the thicknesses of as-deposited NiCr and NiCrSi thin-film resistors as a function of deposition time.

XRD was used to investigate the crystalline properties of the NiCr and NiCrSi thin films at room temperature. All of the XRD patterns (Figure 4) of the NiCr and NiCrSi thin films revealed the amorphous structure, and no crystalline phases were apparently observed, and only the Ag and Al_2O_3 phases were observed (not shown here). These results suggested that the thickness (or deposition time) had no effect on the crystallization of the as-deposited NiCr and NiCrSi thin films. Because the NiCr and NiCrSi thin films were deposited using a sputtering method in a pure Ar atmosphere, we believe that oxidation did not occur during the deposition process. We used FESEM to observe the surface morphologies of the NiCr thin-film resistors (see Figure 5a for deposition time of 10 min and Figure 5b for that of 60 min) and of the NiCrSi thin-film resistors (see Figure 5c for deposition time of 10 min and Figure 5d for that of 60 min). Only nano-crystalline grains were observed, and the surface morphologies were almost unchanged, regardless of the deposition time.

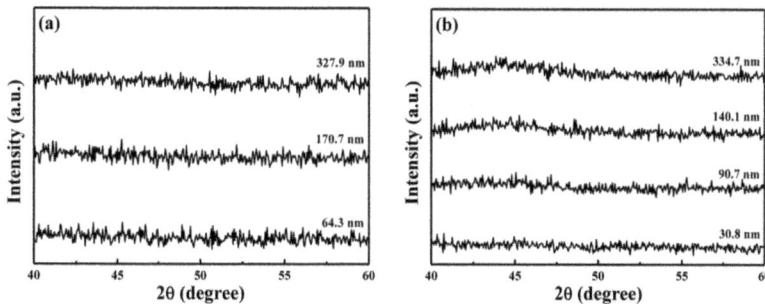

Figure 4. X-ray diffraction (XRD) patterns of 25 °C-deposited (**a**) NiCr and (**b**) NiCrSi thin-film resistors as a function of the thin films' thickness (or deposition time).

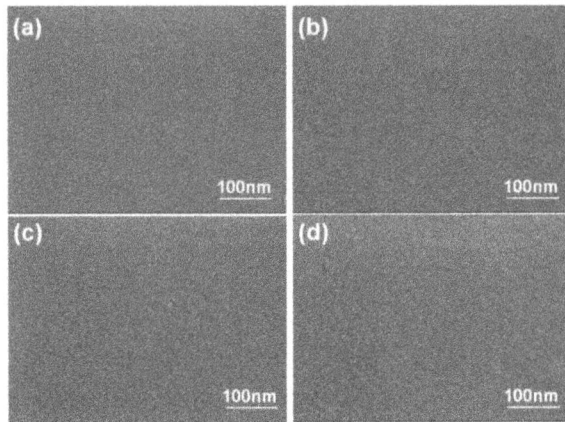

Figure 5. Surface morphologies of 25 °C-deposited thin films as a function of deposition time. NiCr thin films deposited at (**a**) 10 min and (**b**) 60 min. NiCrSi thin films deposited at (**c**) 10 min and (**d**) 150 min.

Figure 6 shows the effects of thickness (deposition time) on resistance and resistivity for NiCr and NiCrSi thin-film resistors measured at 25 °C. The resistances of the NiCr and NiCrSi thin-film resistors were recorded by the four-point probe method, and resistivity was derived from resistance using a measurement of the thin films' thicknesses, shown in Figure 3. As the temperature increased from 25 to 125 °C, the resistance of the NiCr thin-film resistors slightly decreased and that of the NiCrSi thin-film resistors slightly increased. The NiCr and NiCrSi thin-film resistance values at 25 and 125 °C were similar. Figure 6a shows that the resistance of the NiCr thin-film resistors monotonously decreased as the thin films' thickness increased. Figure 6b also shows that the thinner NiCrSi thin-film resistors showed higher resistance, and the resistance reached a saturation value as the thin films' thickness became equal to or greater than 140.1 nm (*i.e.*, when the deposition time was equal to or greater than 60 min). If we suppose that the thicknesses of the NiCr and NiCrSi thin-film resistors were independent of the measured temperature, then the resistivity of the NiCr and NiCrSi thin-film resistors at 25 and 125 °C were similar and the variations in resistivity were not apparently observed.

In the past, only a few papers have discussed the effects of thin films' thickness on their resistance and resistivity, as it has been difficult to discern a correlation between these values. In the free-electron model of metallic thin-film resistors with hard-wall boundary conditions, the discretization of energy levels makes it impossible to treat both the Fermi energy and the electron density as independent of thickness [14]. In this study, as the thin films' thicknesses increased, the log(resistivity) of the NiCr (Figure 6a) and NiCrSi (Figure 6b) thin-film resistors linearly decreased. Katumba and Olumekor found that the $\log(\rho)$ of Cu-MgF_2 cermet thin-film resistors' thickness exhibited an approximately linear decrease from about 110 to 300 nm. Ultimately, they found the relationship between resistivity and thickness for thin-film Cu-MgF_2 cermets to be as follows [2]:

$$\rho_f = \rho_o \times \exp(10 \times S/t) \tag{2}$$

where ρ_f is the resistivity of thin films, ρ_o is the limiting resistivity of very thick cermets, t is the film thickness, and S is a measure of the separation between the metallic islands embedded in the insulator matrix of the cermets. The present study not only sought the qualitative effects of the thicknesses of the NiCr and NiCrSi thin-film resistors but also attempted to quantify the relationships between resistivity and thickness. We found that the log(resistivity) values of the NiCr and NiCrSi thin-film resistors in Figure 6 decreased in an approximately linear mode as their thicknesses increased—similar to Cu-MgF_2 cermet thin-film resistors.

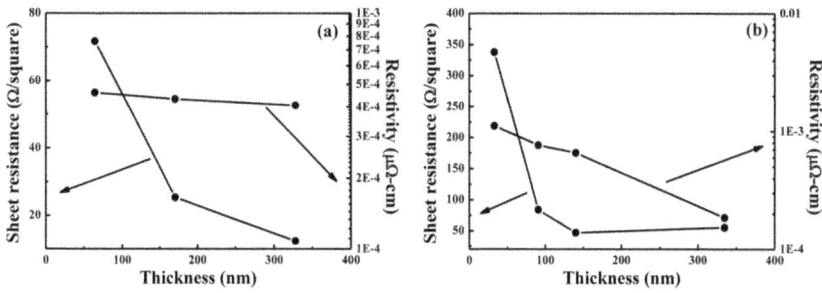

Figure 6. Variations in the resistance and resistivity of 25 °C-measured as-deposited (**a**) NiCr and (**b**) NiCrSi thin-film resistors as a function of the thin films' thickness.

Resistances for materials at any temperature other than standard temperature (usually taken to be 20 °C) on the specific resistance can be determined using the following formula:

$$R = R_{ref} \left[1 + \alpha \left(T - T_{ref} \right) \right] \tag{3}$$

where R is the material resistance at temperature T; R_{ref} is the material resistance at temperature T_{ref}, usually 20 °C or 0 °C; α is the TCR value for the material, symbolizing the resistance change factor per degree of temperature change; T is the material temperature in degrees Celsius; and T_{ref} is the reference temperature that α is specified at for the material.

The TCR values of as-deposited NiCr and NiCrSi thin-film resistors are shown in Figure 7 as a function of the thin films' thicknesses, using the measured results shown in Figure 6. For most pure metals, these TCR values are positive. Dhere *et al.* found that the NiCr thin-film resistors with low positive TCRs (*i.e.*, fewer than 100 ppm/°C) had been obtained at all thicknesses studied when the total atomic content of chromium, oxygen, and carbon reached 50%–55% [15]. The TCR value of Ni metal is 0.00017 and of Cr metal is 13×10^{-8}, meaning that resistance increases with increasing measured temperature. Nevertheless, the TCR value of as-deposited NiCr thin-film resistors was positive in the range of 197.2 to 230.1 ppm/°C, which are larger than those of Ni and Cr metals. Single crystalline Si has a TCR value of about −0.04 (depending strongly on the presence of impurities in the material), so Si could be added to change the TCR value of NiCr-based thin-film resistors. As Figure 7 shows, the TCR values of as-deposited NiCrSi thin-film resistors were negative, in the range of −106.4 to −153.3 ppm/°C, meaning that the resistance decreased as the measured temperature increased. The TCR value of the NiCrSi thin-film resistors, shown in Figure 7, had no significant change as the thin-film's thickness increased from 30.8 nm to 334.7 nm. Ni and Cr are metals, and Si is a semiconductor, and the XRD patterns in Figure 4 show that the NiCrSi thin films were in the amorphous phase. Those results suggest that as the NiCrSi thin films are deposited, the Ni and Cr will form alloys and then NiCr alloys will form a NiCrSi compound with Si. The electrical properties of a compound are the sum total of each component in it. Hence, the electrical properties of Si would affect the TCR value of the NiCrSi thin-film resistors. Ni and Cr, as well as NiCr thin-film resistors, have the positive TCR values. We believe that the negative TCR value of the NiCrSi thin-film resistors was caused by the addition of Si into the NiCr alloy.

As Figure 6 shows, the resistance of NiCr thin-film resistors linearly decreased as their thickness increased with deposition time, and the resistance of the NiCrSi thin-film resistors was almost unchanged as the thickness became equal to or greater than 140.1 nm (10 min deposition time). To simplify the fabrication process, the thickness (deposition time) of the NiCr thin films was fixed at 64.3 nm (10 min), and the thickness (deposition time) of the NiCrSi thin films was set at 30.8 nm (10 min), 90.7 nm (30 min), and 140.1 nm (60 min), respectively. Cross-section images of the as-deposited bi-layer thin-film resistors with their various structures and with the different NiCrSi thin films' thicknesses are presented in Figure 8, where the bi-layer structure is easily observed. Figure 8 shows that the thickness of the NiCr thin films was in the range of 65.5–67.0 nm, which is similar to the value obtained from the single-layer thin films shown in Figure 3. Whether it was being used as the upper layer or the lower layer, it almost had the same value.

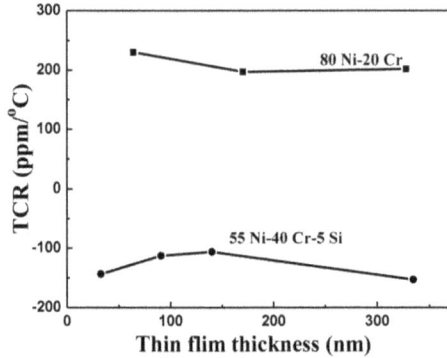

Figure 7. Variations in the temperature coefficient of resistance of as-deposited NiCr and NiCrSi thin-film resistors as a function of the thin-films' thickness.

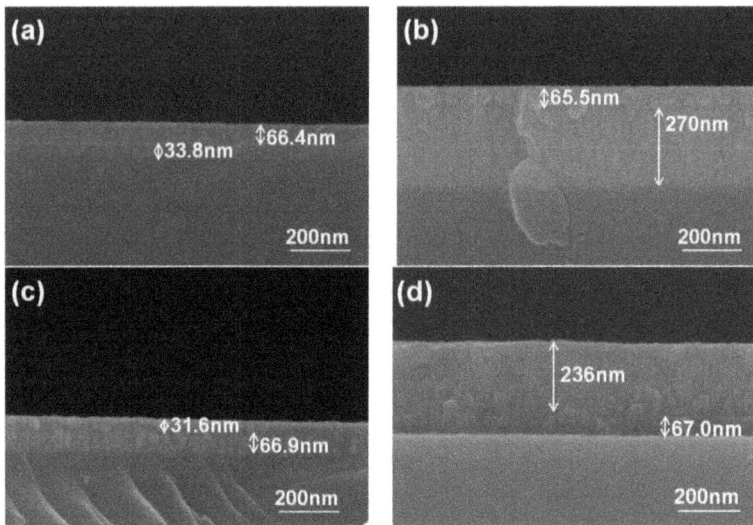

Figure 8. Cross-sectional images of the bi-layer thin-film resistors created using different deposition time of the NiCrSi thin films. In (**a,b**), a NiCr thin film deposited for 10 min was used as the upper layer, and a NiCrSi thin film deposited for (**a**) 10 min or (**b**) 60 min was used as the lower layer. In (**c,d**), a NiCr thin film deposited for 10 min was used as the lower layer, and a NiCrSi deposited thin film deposited for (**c**) 10 min or (**d**) 60 min was the upper layer.

The results in Figure 8 show that the thickness of the NiCrSi thin films in the bi-layer structure also increased with the increase of deposition time, but the thin films had different deposition rates, depending on whether they were used as upper or lower layers. We also investigated how the deposition time of the NiCrSi thin films would affect the thicknesses of the two NiCr-NiCrSi bi-layer structures' thin films, and the results are shown in Figure 9. When the NiCr thin film deposited for

10 min was used as the upper layer and the NiCrSi thin films deposited for 10, 30, or 60 min were used as lower layer, the thickness of the bi-layer thin-film resistors (or the NiCrSi thin films) was about 100.2 nm (33.8 nm), 192.2 nm (125.9 nm), and 335.5 nm (270 nm). When the NiCr thin film deposited for 10 min was used as the lower layer and the NiCrSi thin films deposited for 10, 30, or 60 min were used as upper layer, the thickness of the NiCrSi thin films deposited for 10, 30, or 60 min (or of the NiCrSi thin films) was about 98.5 nm (31.6 nm), 166.6 nm (100.8 nm), and 303 nm (236 nm). We expected that as the deposition time of the NiCrSi thin films increased, so too would the thickness of the bi-layer structures' thin films. Our results also suggest that when the NiCr thin films were used as the lower layer, the NiCrSi thin films had a lower deposition rate.

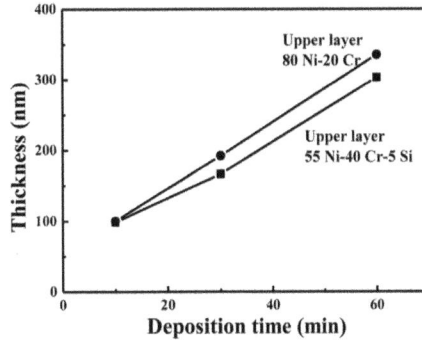

Figure 9. Variations in the thickness of the bi-layer structure as a function of deposition time of the NiCrSi thin films.

We also recorded the bi-layer thin-film resistors' resistance using the four-point probe method and derived the resistivity from the resistance using a measurement of the thin films' thicknesses, as shown in Figure 9. In addition, we examined the effect of the NiCrSi deposition time on the resistance and resistivity of the bi-layer thin-film resistors as a function of the NiCrSi thin films' thickness, as Figure 10 shows. The results in Figure 10 indicate two important points: (i) regardless of whether the NiCrSi thin films were used as the upper or the lower layer, the resistance of the bi-layer structure decreased as the NiCrSi thin films' thickness (or the deposition time) increased; (ii) the resistivity of the bi-layer structure remained stable even if the NiCrSi thin films' thickness increased. These results suggest that thin-film resistors with stable resistances can easily be achieved by using a bi-layer structure.

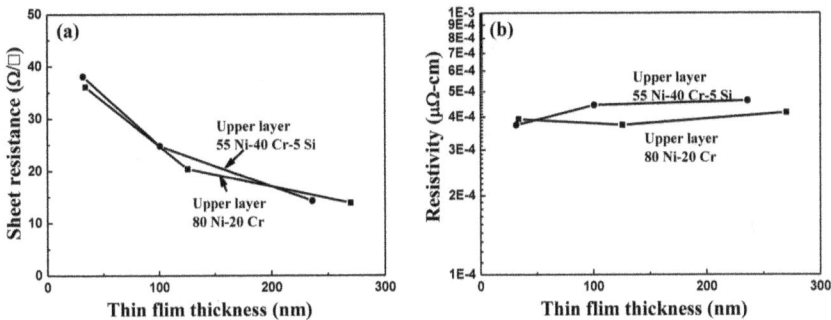

Figure 10. Variations in the (**a**) resistance and (**b**) resistivity of the bi-layer thin-film resistors as a function of the NiCrSi thin films' thickness.

Figure 11 presents the TCR values of our bi-layer thin-film resistors as a function of the NiCrSi thin films' thickness. We found that the deposition time of the NiCrSi thin films in the two different structures had a large effect on the TCR values of the NiCr-NiCrSi bi-layer thin-film resistors. When the NiCrSi thin films were used as the upper layer and their thickness increased from 31.6 nm to 236 nm, the TCR value of the bi-layer thin-film resistors dropped from 118.1 to 35.1 ppm/°C, coming close to zero ppm/°C as the NiCrSi thin films' thickness increased. When the upper layer was the NiCr thin films and the thickness of NiCrSi thin films increased from 33.8 to 270 nm, the TCR value of the bi-layer thin-film resistors shifted from 110.8 to −72.4 ppm/°C. Hence, the TCR changed from close to 0 ppm/°C to a negative value as the NiCrSi thin films' thickness increased. Compared with the thickness of the bi-layer structure shown in Figure 9, when the NiCrSi thin films were used as the lower (upper) layer, their thickness increased to 33.8 (31.6) nm, 125.9 (100.8) nm, and 270 (236) nm, respectively.

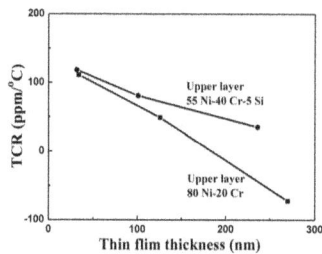

Figure 11. Temperature coefficient of resistance of bi-layer thin-film resistors as a function of our NiCrSi thin films' thickness.

Except the effect of the thickness of the NiCrSi thin films, the material in contact with the Ag electrode is another possible factor affecting the TCR value in these bi-layer thin-film resistors. Many scattering effects are believed to influence the resistivity of bi-layer thin-film resistors, including the surface scattering effect, the grain boundaries (or interface) scattering effect, the uneven or rough surfaces scattering effect, and the impurities scattering effect [16]. In a thin-film material, if, as proposed, the thin films have smooth or even surfaces, then surface scattering is believed to be the main factor affecting their electrical properties. Even the splitting is not really observed in the bi-layer thin films shown Figure 8, the variations of TCR values are apparently influenced by the stacking method and thickness of NiCrSi thin films. If the lower layer is conducted with the Ag electrode, the ohmic conduction mechanism will dominate due to the contact between the lower-layer materials (NiCr or NiCrSi thin films) and the Ag electrode. We believe that an interface layer exists between the upper-layer and lower-layer thin-film materials, so an interface scattering effect and a rough surface scattering effect will happen at the contact boundaries, causing the variations in the TCR values of the bi-layer thin-film resistors. Those results suggest that the thickness of the NiCrSi thin films will dominate the TCR values in such bi-layer thin-film resistors. Those results also suggest that the bi-layer structure is an important technology for developing thin-film resistors with TCR values close to 0 ppm/°C.

4. Conclusions

In our bi-layer structure, regardless of whether NiCr thin films were used as the lower layer or the upper layer, their thickness was around 65 nm. As the deposition time was 10 min, the thickness of the NiCrSi thin films in these bi-layer structures was similar to the thickness in a single-layer structure. Their resistances decreased as the deposition time of the NiCrSi thin films increased. The TCR values of our as-deposited NiCr and NiCrSi thin-film resistors were in the range of 197.2 to 230.1 ppm/°C and −106.4 to −153.3 ppm/°C, respectively. For the bi-layer thin-film resistors, as the deposition time of the NiCrSi thin films increased from 10 to 60 min, the thickness increased from 31.6 to 236 nm when we used them as the upper layer, and from 33.8 to 270 nm when we used NiCr thin films as the upper layer. As the deposition time of the NiCrSi thin films increased from 10 to 60 min, the TCR value changed from 118.1 to 35.1 ppm/$^\circ$C when we used NiCrSi thin films as the upper layer, and from 110.8 to -72.4 ppm/$^\circ$C when we used NiCr thin films as the upper layer.

Acknowledgments: The authors acknowledge financial supports of MOST 104-2221-E-390-013-MY2 and MOST 104-2622-E-390-004-CC3.

Author Contributions: Huan-Yi Cheng, Chi-Lun Li, and Pei-Jou Li helped proceeding the deposition and measurement processes and data analysis; Ying-Chung Chen and Cheng-Fu Yang organized the paper and encouraged in paper writing; Mau-Phon Houng helped proceeding the experimental processes and measurements.

Conflicts of Interest: The authors declare no conflict of interest.

References

1. Mcalister, S.P.; Inglis, A.D.; Kroeker, D.R. Crossover between hopping and tunnelling conduction in Au-SiO$_2$ films. *J. Phys. C* **1984**, *17*, L751–L756. [CrossRef]
2. Katumba, G.; Olumekor, L. Effects of thickness and composition on the resistivity of Cu-MgF$_2$ cermet thin film resistors. *J. Mater. Sci.* **2000**, *35*, 2557–2559. [CrossRef]
3. Lai, L.; Zeng, W.; Fu, X.; Sun, R.; Du, R. Annealing effect on the electrical properties and microstructure of embedded Ni–Cr thin film resistor. *J. Alloys Compd.* **2012**, *538*, 125–130. [CrossRef]
4. Wang, X.Y.; Ma, J.X.; Li, C.G.; Shao, J.Q. Structure and electrical properties of quaternary Cr–Si–Ni–W films prepared by ion beam sputter deposition. *J. Alloys Compd.* **2014**, *604*, 12–19. [CrossRef]
5. Ceccone, G.; Nicholas, M.G.; Peteves, S.D.; Kodentsov, A.A.; Kivilahti, J.K.; Loo, F.J.J. The Brazing of S$_3$N$_4$ with Ni-Cr-Si Alloys. *J. Eur. Ceram. Soc.* **1995**, *15*, 563–572. [CrossRef]
6. Thompson, J.R.; Goyal, A.; Christen, D.K.; Kroeger, D.M. Ni-Cr textured substrates with reduced ferromagnetism for coated conductor applications. *Physica C* **2002**, *370*, 169–176. [CrossRef]
7. Abdul-Razzaq, W.; Amoruso, M. Electron transport properties of Ni and Cr thin films. *Physica B* **1998**, *253*, 47–51. [CrossRef]
8. Nocerinot, G.; Singer, K.E. The Electrical and Compositional Structure of Thin Ni-Cr films. *Thin Solid Films* **1979**, *57*, 343–348. [CrossRef]
9. Vinayak, S.; Vyas, H.P.; Vankar, V.D. Microstructure and electrical characteristics of Ni–Cr thin films. *Thin Solid Films* **2007**, *515*, 7109–7116. [CrossRef]
10. Kotisa, L.; Menyhard, M.; Toth, L.; Zalar, A.; Panjan, P. Determination of relative sputtering yield of Cr/Si. *Vacuum* **2008**, *82*, 178–181. [CrossRef]
11. Hieber, K.; Dittmann, R. Structural and Electrical Properties of CrSi$_2$ Thin Film Resistors. *Thin Solid Films* **1976**, *36*, 357–360. [CrossRef]
12. Zhang, Y.; Dong, X.P.; Wu, J. Microstructure and electrical characteristics of Cr–Si–Ni films deposited on glass and Si(100) substrates by RF magnetron sputtering. *Mater. Sci. Eng. B* **2004**, *113*, 154–160. [CrossRef]
13. Lee, B.J.; Shin, P.K. Fabrication Characterization of Ni-Cr Alloy Thin Films for Application to Precision Thin Film Resistors. *J. Electr. Eng. Technol.* **2007**, *2*, 525–531.
14. Wang, Z.; Wang, S.; Shen, S.; Zhou, S. Impurity resistivity of an ideal metallic thin film. *Phys. Rev. B* **1997**, *55*, 10863–10868. [CrossRef]
15. Dhere, N.G.; Vaiude, D.G.; Losch, W. Composition and temperature coefficient of resistance of Ni-Cr thin films. *Thin Solid Films* **1979**, *59*, 33–41. [CrossRef]
16. Lacy, F. Developing a theoretical relationship between electrical resistivity, temperature, and film thickness for conductors. *Nanoscale Res. Lett.* **2011**, *6*. [CrossRef] [PubMed]

nanomaterials

MDPI

Article

Thermal Plasma Synthesis of Crystalline Gallium Nitride Nanopowder from Gallium Nitrate Hydrate and Melamine

Tae-Hee Kim [1], Sooseok Choi [2,*] and Dong-Wha Park [1,*]

[1] Department of Chemistry and Chemical Engineering and Regional Innovation Center for Environmental Technology of Thermal Plasma (RIC-ETTP), Inha University, 100 Inha-ro, Nam-gu, Incheon 22212, Korea; taehee928@naver.com

[2] Department of Nuclear and Energy Engineering, Jeju National University, 102 Jejudaehak-ro, Jeju 63243, Korea

* Correspondence: sooseok@jejunu.ac.kr (S.C.); dwpark@inha.ac.kr (D.-W.P.); Tel.: +82-64-754-3644 (S.C.); +82-32-860-7468 (D.-W.P.)

Academic Editors: Krasimir Vasilev and Melanie Ramiasa
Received: 30 December 2015; Accepted: 10 February 2016; Published: 24 February 2016

Abstract: Gallium nitride (GaN) nanopowder used as a blue fluorescent material was synthesized by using a direct current (DC) non-transferred arc plasma. Gallium nitrate hydrate ($Ga(NO_3)_3 \cdot xH_2O$) was used as a raw material and NH_3 gas was used as a nitridation source. Additionally, melamine ($C_3H_6N_6$) powder was injected into the plasma flame to prevent the oxidation of gallium to gallium oxide (Ga_2O_3). Argon thermal plasma was applied to synthesize GaN nanopowder. The synthesized GaN nanopowder by thermal plasma has low crystallinity and purity. It was improved to relatively high crystallinity and purity by annealing. The crystallinity is enhanced by the thermal treatment and the purity was increased by the elimination of residual $C_3H_6N_6$. The combined process of thermal plasma and annealing was appropriate for synthesizing crystalline GaN nanopowder. The annealing process after the plasma synthesis of GaN nanopowder eliminated residual contamination and enhanced the crystallinity of GaN nanopowder. As a result, crystalline GaN nanopowder which has an average particle size of 30 nm was synthesized by the combination of thermal plasma treatment and annealing.

Keywords: thermal plasma; annealing; gallium nitride; gallium nitrate hydrate; melamine; nanopowder

1. Introduction

Gallium nitride (GaN) has been used as a binary III–V direct band-gap semiconductor material in light-emitting diodes since the 1990s. GaN is a blue fluorescence material used in LEDs. Lighting devices create various colors by combining red (R), green (G), and blue (B) [1]. In order to prepare a white light source, blue fluorescence is mixed with other light sources such as yellow, red, or green light. Such white LEDs are taking the place of traditional incandescent and fluorescent lights. GaN has a large band gap energy of 3.4 eV at room temperature and a high thermal conductivity of 130 W/m·K [2–4]. It is possible to use these materials in optoelectronic devices which have wide band gap with energies from the visible to the deep ultraviolet region. Gallium nitride is a very hard material that has a strong atomic bonding as a wurtzite crystal structure. It can be used for applications in optoelectronic, high-power, high-frequency, and high temperature devices. For example, GaN can be applied as the substrate which makes violet laser diodes at 405 nm without the requirement of nonlinear optical frequency-doubling. It is usually applied by deposition on silicon carbide (SiC) and sapphire (Al$_2$O$_3$) plates. During the deposition of GaN onto plates, the mismatching of GaN lattice structure could occur, and doping to n-type and p-type materials with silicon or oxygen elements has appeared [5,6]. This mismatch disturbs the crystal growth and leads to defects of crystal GaN due to increased tensile stress. In order to produce an excellent GaN device, crystalline GaN should be grown uniformly on the plate.

In the present industry, gallium nitride is often grown on foreign substrates on thin films by MOVPE (metal organic vapor phase epitaxy) and MOCVD (metal organic chemical vapor deposition). However, the thermal expansion coefficient of GaN is considerably higher than that of silicon or sapphire. These different properties invite a crack of epitaxial films or wafer bowing in the GaN growth process on the substrate. Therefore, it is difficult to produce bulk single crystals.

GaN powder can be produced by a novel hot mechanical alloying process. It requires a lengthy process to yield the powder by this method and the purity of the product is not sufficient [7]. In the methods for synthesis of GaN powder, GaN can be synthesized from molten gallium metal with a stream of reactive NH$_3$ gas in a furnace. Gallium metal is maintained at a molten state at 30 °C. However, it is difficult to vaporize at a low temperature due to its high vaporization temperature of 2400 °C. Although this method is simple and economic, it is not suitable to produce quality GaN powder [2,8]. The ammonothermal reduction nitridation method can produce GaN powder by reacting gallium oxide with NH$_3$ in the temperature range between 600 and 1100 °C. However, incomplete nitridation of the oxide easily occurs. In other words, the purity of synthesized GaN is fairly low. Gallium phosphide (GaP) and Gallium arsenide (GaAs) are possible alternative materials for Gallium oxide in the temperature range of 1000–1100 °C [2,9]. Carbothermal reduction nitridation is applied to synthesize GaN nanorods and nanowires. Carbon is employed as a catalyst for crystal growth [3]. Therefore, an additional decarbonization process is necessary to eliminate the residual carbon. Liquid precursors can be used by an aerosol-assisted vapor phase synthesis method. It is completed at a relatively low temperature. However, this method has a multitude of steps such as preparation of gallium solid compounds, oxidation of gallium precursor, and nitridation chemistry. Therefore, it takes an extended synthesis time to produce GaN powder [10]. Arc plasma has been regarded as an effective method to obtain nanometer-sized GaN. This method requires a very short time compared with other synthesis methods [11,12]. However, gallium lump is an expensive precursor and its evaporation requires an extended time in the arc plasma region. Generally, conventional synthesis methods have several limitations for the preparation of nano-sized particles and complete nitridation.

GaN has commonly been grown in the industrial field by the epitaxial method on the substrate. However, it is expected that the uniform deposition of GaN is achievable using the nanoparticle printing method rather than the conventional chemical vapor deposition (CVD) method [6]. Therefore, using the new method would mean that the fine particles with a high crystallinity could be printed on a substrate. It was expected that this method would prevent the mismatch of lattice and irregular growth. In this work, thermal plasma which is able to produce GaN fine particles was applied as substitute technology of the conventional CVD method. Gallium nitrate hydrate (Ga(NO$_3$)$_3$·xH$_2$O) was used as the raw material instead of Ga or Ga$_2$O$_3$ which have been used in previous studies to synthesize GaN [2–4,10,13–20]. However, the raw material itself has abundant oxygen elements. Therefore, melamine (C$_3$H$_6$N$_6$) was additionally injected into the thermal plasma jet to prevent the oxidation of decomposed Ga into Ga$_2$O$_3$ [21,22]. In order to improve the crystallinity of synthesized GaN, products from the thermal plasma were subjected to annealing using a vacuum furnace. The crystallinity and purity of synthesized GaN nanopowder were investigated before and after annealing.

2. Experimental Setup

GaN nanopowder was synthesized from Ga(NO$_3$)$_3$·xH$_2$O (99.9% purity, Alfa Aeser Inc., Boston, MA, USA) and C$_3$H$_6$N$_6$ (99% purity, Ald rich Inc., St. Louis, MO, USA) using the non-transferred DC arc plasma. The schematic diagram of the thermal plasma system is indicated in Figure 1. The system consists of a DC power supply (YC-500TSPT5, Technoserve, Toyohashi, Japan), a plasma torch (SPG-30N2S, Technoserve, Japan), a powder feeder (ME-14C, SHINKO Electric Co Nagoya, Japan) for the injection of the precursor, a chamber, and a crucible for the precursor. The thermal plasma jet was generated in the plasma torch by Ar or Ar–N_2, which formed gases under atmospheric pressure. The thermal plasma jet was ejected from a 6 mm diameter nozzle by the thermal expansion of plasma, forming gas by an electric arc channel which was connected between a conical cathode and the anode nozzle. $Ga(NO_3)_3 \cdot xH_2O$ powder was used as the raw material for the synthesis of GaN nanopowder. The injected precursor powder was hundreds of micrometers in size, and its morphology is not specific. The precursor was a pellet which had a diameter of 45 mm and a thickness of 10 mm created by a hydraulic press. The precursor pellet was placed on a tungsten crucible which was fixed by a water-cooling holder. As a result, the precursor was rapidly melted and evaporated by the confronting thermal plasma jet, as shown in Figure 1. The vaporization temperature of $Ga(NO_3)_3 \cdot xH_2O$ is lower than 100 °C. Therefore, it is a more economical precursor than Ga or Ga_2O_3 which were used as precursors to synthesize GaN in previous work [23]. However, it contains excessive nitrogen, oxygen, and hydrogen elements. The complete dissociation of the oxygen from $Ga(NO_3)_3 \cdot xH_2O$ is difficult due to the high latent heat of H_2O. For this reason, it is normal that evaporated $Ga(NO_3)_3 \cdot xH_2O$ is oxidized to Ga_2O_3 by its inherent oxygen elements. Therefore, melamine ($C_3H_6N_6$) powder was used to prevent the production of gallium oxide. The injected $C_3H_6N_6$ was converted to carbon, nitrogen, hydrogen, and other molecules by decomposition at high temperatures in the thermal plasma jet. These byproduct molecules consume the oxygen elements of the $Ga(NO_3)_3 \cdot xH_2O$ precursor. Injected melamine powder consists of nonspecific-shaped particles under 500 nm. In addition, ammonia (NH_3) gas was diagonally injected into the pellet of $Ga(NO_3)_3 \cdot xH_2O$ from the upper side of the chamber to nitride the Ga element.

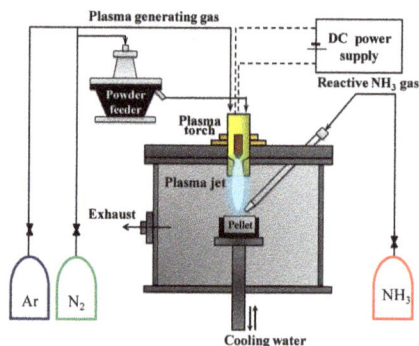

Figure 1. Experimental apparatus for the synthesis of GaN (gallium nitride) nanopowder by DC (direct current) non-transferred thermal plasma.

The detailed operating conditions are indicated in Table 1. In the first case, the condition of Plasma 1, the $Ga(NO_3)_3 \cdot xH_2O$ pellet was reacted with only NH_3 gas and without $C_3H_6N_6$. The thermal plasma jet was generated by mixed argon and nitrogen gases. The high thermal conductive nitrogen was injected as a plasma forming gas to help the nitridation or evaporation as the input power was increased. The plasma input power was 12.6 kW at the fixed current of 300 A and the average voltage of 42 V. In the other cases of Plasma 2, 3, 4, and 5, the $Ga(NO_3)_3 \cdot xH_2O$ pellet was evaporated by pure Ar thermal plasma together with $C_3H_6N_6$. In these cases, the input power for the generation of the thermal plasma jet was 8.4 kW with the current fixed at 300 A and an average voltage of 28 V. Argon is a monoatomic molecule and nitrogen is a diatomic molecule. In order to generate thermal plasma from nitrogen gas, more energy is required compared to argon gas. Accordingly, electric resistance for arc generation between the cathode and the anode is increased. Therefore, argon thermal plasma has a lower average voltage with lower resistance than argon–nitrogen thermal plasma at the fixed current. Input power of 8.4 kW was sufficient to vaporize the $Ga(NO_3)_3 \cdot xH_2O$ pellet due to its low vaporization temperature of 80 °C. Synthesized nanoparticles were attached on the inner surface of the reactor which was chilled by cooling water. The product was collected by scraping with a thin film for the post processing and analysis.

Table 1. Operating conditions for the synthesis of GaN powder by thermal plasma.

Experiment No.		Plasma 1	Plasma 2	Plasma 3	Plasma 4	Plasma 5
Condition of precursors	Weight of $Ga(NO_3)_3 \cdot xH_2O$			6 g		
	Weight of $C_3H_6N_6$	-	15 g (pellet)	15 g (powder injection)	8.8 g	15 g (powder injection)
	Molar ratio of $Ga(NO_3)_3 \cdot xH_2O$ and $C_3H_6N_6$	-	1:6	1:6	1:3	1:6
Condition of plasma generating	Flow rate of plasma forming gas	13 L/min Ar + 2 L/min N_2		13 L/min Ar		
	Flow rate of reactive NH_3 gas	10 L/min	3 L/min	3 L/min	3 L/min	-
	Flow rate of carrier gas	-	-	3 L/min N_2	3 L/min N_2	3 L/min N_2
	Plasma input power	12.6 kW (300 A, 42 V)		8.4 kW (300 A, 28 V)		

$C_3H_6N_6$ powder was used in the three different procedures in Plasma 2, 3, 4, and 5. A detailed illustration of $C_3H_6N_6$ injection methods is indicated in Figure 2a–d. $C_3H_6N_6$ was mixed with $Ga(NO_3)_3 \cdot xH_2O$, they were pressed as the pellet precursor in Plasma 2. $C_3H_6N_6$ powder was injected into the high temperature of the thermal plasma jet as powder through the anode electrode in Plasmas 3, 4, and 5. The molar ratio of $Ga(NO_3)_3 \cdot xH_2O$ and $C_3H_6N_6$ was controlled at 1:6 and 1:3 in Plasmas 3, 4, and 5. In these cases, NH_3 gas was injected into the pellet as a nitridation source by the probe as in Figure 1. $Ga(NO_3)_3 \cdot xH_2O$ was reacted with only $C_3H_6N_6$ powder and without NH_3 gas in Plasma 5, in order to check the possibility of nitridation by the nitrogen element of $C_3H_6N_6$. The products from thermal plasma were annealed in a vacuum furnace (SH-TMFGF-50, Samheung Inc., Sejong, Korea) to eliminate contamination and to enhance the crystallinity. It was carried out at 850 °C for three hours.

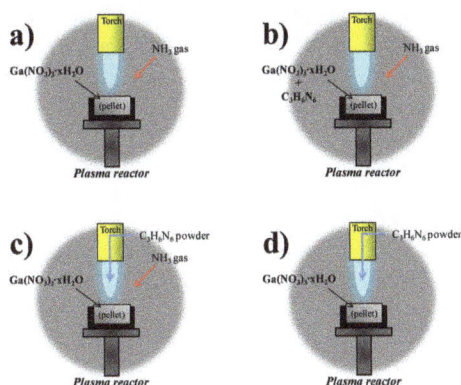

Figure 2. Detailed illustration of $C_3H_6N_6$ injection methods in Plasma 1, 2, 3, 4, and 5; (**a**) Plasma 1; (**b**) Plasma 2; (**c**) Plasmas 3 and 4; (**d**) Plasma 5.

A TGA analysis result of $Ga(NO_3)_3 \cdot xH_2O$ as raw materials according to temperature is indicated in Figure 3. Analysis was conducted from 40 °C to 800 °C with an increase in increments of 10 °C/min under a nitrogen atmosphere. An obvious mass reduction of approximately 67% occurred between 60 °C and 180 °C as is shown in Figure 3. $Ga(NO_3)_3 \cdot xH_2O$ was converted to Ga_2O_3, H_2O, and N_2O_3 gas with increasing temperatures, which is in accordance with previous studies [23]. Therefore, it can be assumed that the final product of Ga_2O_3 with a mass reduction of 78% would be produced at 800 °C. As a result, the estimated value of x in $Ga(NO_3)_3 \cdot xH_2O$ was determined to be 32.1 based on the mass reduction from 12.45 mg at 40 °C to 2.80 mg at 800 °C. Accordingly, $Ga(NO_3)_3 \cdot xH_2O$ is referred to as $Ga(NO_3)_3 \cdot 32H_2O$ in this work.

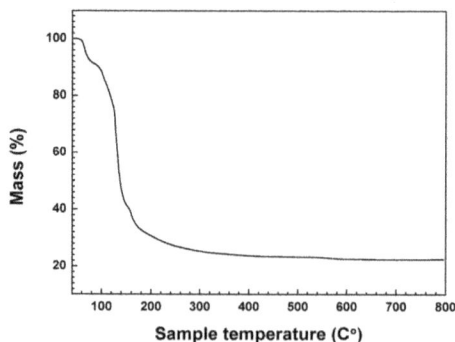

Figure 3. TGA (thermogravimetric analysis) of $Ga(NO_3)_3 \cdot xH_2O$ raw material.

The crystallinity of the synthesized powder was observed by XRD (X-ray diffraction, DMAX 2500, Rigaku Co., Akishima, Japan) with Cu Kα source. Morphology and particle size were analyzed by FE-SEM (Field-emission scanning electron microscopy, S-4300, Hitachi Co., Tokyo, Japan) and FE-TEM (Field-emission transmission electron microscopy, JEM-2100F, Jeol, Japan). In addition, the elemental composition of particles was confirmed by an EDS (Energy dispersive spectroscopy) with SEM and TEM. The mean particles size was calculated from the variation of the hundreds of different synthesized particles in the FE-SEM images. TGA (thermogravimetric analysis, Diamond TG-DTA Lab system, Perkin Elmer, Waltham, MA, USA) was applied to investigate the atomic ratio of $Ga(NO_3)_3 \cdot xH_2O$ as raw materials and the thermal behavior of particles synthesized by the thermal plasma and annealing procedure. Chemical bonding and nanostructure of particles synthesized by the thermal plasma and vacuum furnace method were analyzed by XPS (X-ray photoelectron spectroscopy, K-Alpha, Thermo scientific Inc., Waltham, MA, USA).

3. Results and Discussion

Figure 4a shows the thermodynamic equilibrium composition of melamine at changes in temperature from 500 to 6000 K. The calculation was conducted by a commercial software of FactSage (Ver. 6.4, CRCT&GTT, Canada and Germany). The purpose of thermodynamic equilibrium calculation was to expect a preferable chemical reaction and stable chemical species in a given temperature range. $C_3H_6N_6$ containing carbon, hydrogen, and nitrogen elements can be decomposed and dissociated at about 350 °C. The melamine is converted to cyanides, hydrocarbons, and carbon in the high temperature plasma zone. These decomposed or dissociated $C_3H_6N_6$ byproducts can react with the oxygen of a $Ga(NO_3)_3 \cdot 32H_2O$ precursor. Furthermore, the various exothermic gases such as CO, CO_2, H_2, N_2, NO_2, and the releasing of heat generated by the oxygen capture reaction with $Ga(NO_3)_3 \cdot 32H_2O$ and $C_3H_6N_6$ promote the synthesis of GaN nanopowder.

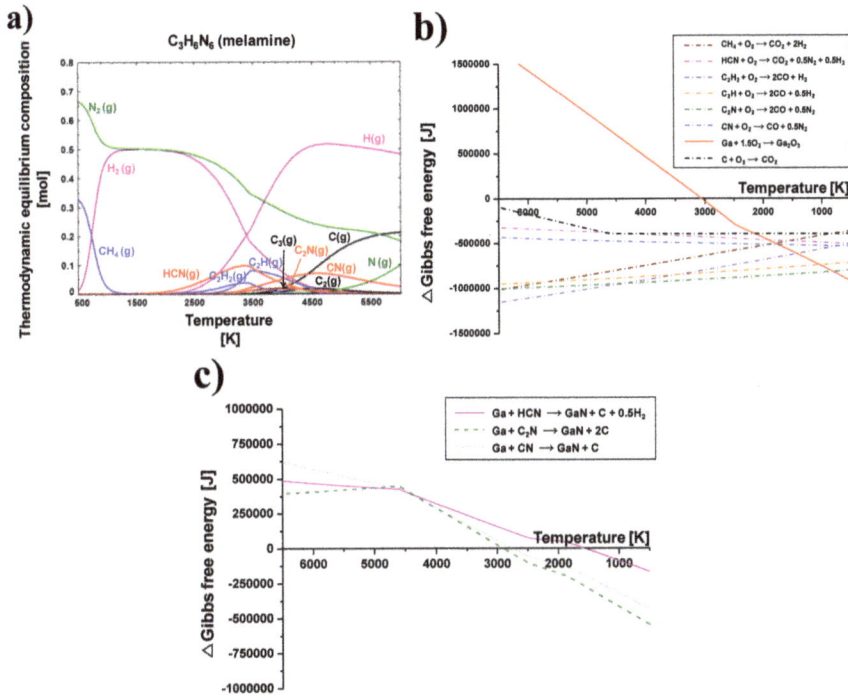

Figure 4. Thermodynamic equilibrium calculation at changes in temperature; (**a**) thermodynamic equilibrium composition of melamine; (**b**) change in Gibbs free energy during the oxygen capture reaction of $C_3H_6N_6$ from $Ga(NO_3)_3 \cdot xH_2O$; and (**c**) change in Gibbs free energy of nitridation reaction by decomposed $C_3H_6N_6$.

The change in Gibbs free energy during oxygen capture reactions with byproducts converted by decomposition of the $C_3H_6N_6$ which was injected into the thermal plasma jet are shown in Figure 4b. The graph consists of seven dotted lines and one solid line. The dotted lines indicate the oxidation by byproducts generated from $C_3H_6N_6$ decomposition, while the solid line indicates the oxidation of gallium to Ga_2O_3. Oxidation reactions of hydrocarbons, cyanides, and carbon generated by the decomposition of $C_3H_6N_6$ are thermodynamically spontaneous throughout the whole temperature range because Gibbs free energies are less than zero. Conversely, Ga_2O_3 can be produced spontaneously under 3100 K. However, oxidation of byproducts is still predominant at temperatures above 750 K owing to a greater amount of negative Gibbs free energy. Therefore, $C_3H_6N_6$ powder is suitable to capture oxygen molecules of $Ga(NO_3)_3 \cdot 32H_2O$ before they oxidize with gallium, even though various byproducts could be produced from the decomposition of $C_3H_6N_6$. For this reason, $C_3H_6N_6$ powder was injected into the thermal plasma jet to facilitate the synthesis of GaN nanopowder. Moreover, the TGA analysis and thermodynamic equilibrium calculation revealed that six moles of $C_3H_6N_6$ powder were needed, while only one mole of $Ga(NO_3)_3 \cdot 32H_2O$ was required to sufficiently convert oxygen from $Ga(NO_3)_3 \cdot 32H_2O$ into carbon oxides, nitrogen oxides, and hydrogen oxides.

Nitridation by the nitrogen element of $C_3H_6N_6$ is considered in Figure 4c. The thermodynamic equilibrium is calculated for the nitridation reaction by HCN, C_2N, and CN molecules. These components containing nitrogen elements were reacted with the Ga element at temperatures under 3000 K. Although ΔG of the three nitridation reactions have negative values, they are lower than those of the oxidation reaction in Figure 4b. ΔG values for the oxidation reaction of HCN, C_2N, and CN at about 3000 K are under 500,000 J, as shown in Figure 4b. Those values for their nitridation reaction at approximately 3000 K are nearly zero. In other words, oxidation reaction is absolutely superior compared to the nitridation reaction at all temperature ranges. Therefore, the byproducts, which are converted from the decomposition of $C_3H_6N_6$, usually react with the oxygen elements of the $Ga(NO_3)_3 \cdot 32H_2O$ raw material.

XRD patterns of products synthesized in Plasma 1 and 2 conditions are shown in Figure 5a,b, respectively. For Plasma 1, only a $Ga(NO_3)_3 \cdot 32H_2O$ pellet was used with 10 L/min NH_3 to synthesize GaN nanopowder without $C_3H_6N_6$. The plasma jet was generated by argon and nitrogen mixed gas in a conventional synthesis process of nitride materials by thermal plasma. All peaks shown in Figure 5a correspond to Ga_2O_3. The results revealed that $Ga(NO_3)_3 \cdot 32H_2O$ is not nitrided solely by NH_3 gas, which is a typical nitrogen source for the nitridation reaction. Ionized nitrogen gas which is produced by the generating thermal plasma jet was not suitable for the nitridation of $Ga(NO_3)_3 \cdot 32H_2O$. $Ga(NO_3)_3 \cdot 32H_2O$ has a low evaporation temperature and is rapidly vaporized in the high temperature environment of the thermal plasma jet. However, nitridation of Ga did not occur, and oxidation of Ga was completed by the presence of abundant oxygen elements from the precursor itself. NH_3 gas usually decomposes into N_2 and H_2 gases at high temperatures. As shown in the following chemical reactions, the oxidation of Ga dominates over the reaction with N_2 and NH_3 to GaN.

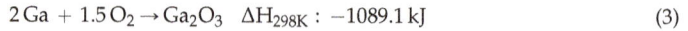

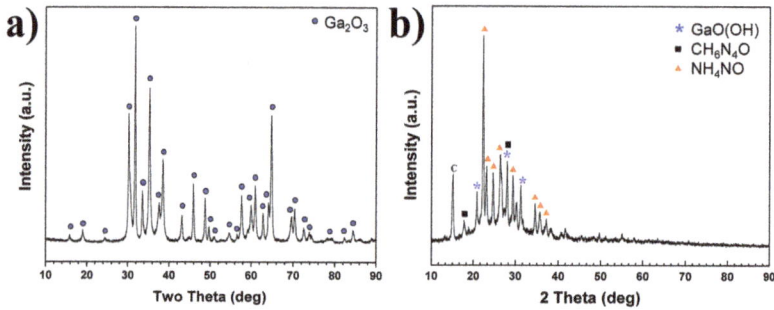

$$Ga + 0.5\,N_2 \rightarrow GaN \quad \Delta H_{298K} : -109.6\,kJ \tag{1}$$

$$Ga + NH_3 \rightarrow GaN + 1.5\,H_2 \quad \Delta H_{298K} : -63.7\,kJ \tag{2}$$

$$2\,Ga + 1.5\,O_2 \rightarrow Ga_2O_3 \quad \Delta H_{298K} : -1089.1\,kJ \tag{3}$$

Figure 5. XRD (X-ray diffraction) pattern of product synthesized by thermal plasma; (**a**) Plasma 1 by $Ga(NO_3)_3 \cdot 32H_2O$ and NH_3 without $C_3H_6N_6$; (**b**) Plasma 2 by $Ga(NO_3)_3 \cdot 32H_2O$ and $C_3H_6N_6$ pellet with NH_3 gas.

These findings indicate that the nitridation of $Ga(NO_3)_3 \cdot 32H_2O$ is difficult without a substance that can consume oxygen as a reductant. Therefore, $C_3H_6N_6$ powder was used as a reductant to oxidize with the abundant oxygen elements of the $Ga(NO_3)_3 \cdot 32H_2O$ precursor.

$C_3H_6N_6$ powder was used as a pellet by compressing it with $Ga(NO_3)_3 \cdot 32H_2O$ as a precursor in Plasma 2. The pellet consisting of $Ga(NO_3)_3 \cdot 32H_2O$ and $C_3H_6N_6$ was vaporized by a thermal plasma jet. NH_3 gas was injected into the pellet at 3 L/min. The XRD pattern of synthesized product at Plasma 2 conditions is indicated in Figure 5b. The peaks are analyzed as gallium hydroxide (GaO(OH)), carbohydrazide (CH_6N_4O), and ammonium nitrate (NH_4NO). The vaporized Ga elements were oxidized, and $C_3H_6N_6$ was not applied in sufficient amounts to act as a reductant. Unlike the experiment involving $Ga(NO_3)_3 \cdot 32H_2O$ and NH_3 gas without $C_3H_6N_6$, GaO(OH) was produced. Products including C-H-O bonding or ammonium ions were produced in the experiment using a pellet mixed with $C_3H_6N_6$ and $Ga(NO_3)_3 \cdot 32H_2O$. It was assumed that $C_3H_6N_6$ could not be used to capture oxygen when it was mixed with $Ga(NO_3)_3 \cdot 32H_2O$ in a pellet. Although $C_3H_6N_6$ affects the product, the effect is not sufficient to nitrate the gallium. $Ga(NO_3)_3 \cdot 32H_2O$ was vaporized above 80 °C. The vaporizing temperature of $C_3H_6N_6$ is 345 °C. The difference of vaporization temperatures for the two raw materials should be considered to prevent oxidation of the Ga element. The useful reductants generated by vaporization of $C_3H_6N_6$ have to react with the gallium element before oxidation. In order to vaporize $C_3H_6N_6$ early, it was injected into the higher temperature region of the thermal plasma.

$C_3H_6N_6$ powder was fed solely into a high temperature thermal plasma jet through two nozzles inside the anode electrode to enable more active decomposition. $Ga(NO_3)_3 \cdot 32H_2O$ has an evaporation temperature below 100 °C, and rapidly vaporized when it contacted the high temperature thermal plasma jet. The molar ratio of $Ga(NO_3)_3 \cdot 32H_2O$ and $C_3H_6N_6$ was controlled at 1:6 and 1:3 in Plasma 3 and 4 conditions. Initially, $C_3H_6N_6$ powder was injected at a 1:6 molar ratio according to the calculation from the TGA analysis of $Ga(NO_3)_3 \cdot 32H_2O$ raw materials. This molar ratio was sufficient to oxidize $C_3H_6N_6$ from the oxygen element of the $Ga(NO_3)_3 \cdot 32H_2O$. The graph in Figure 6 shows XRD patterns of the product synthesized in Plasma 3. The main peaks correspond to $C_3H_6N_6$ and other peaks reflect $C_xH_yN_z$ bound byproducts such as melam, melem and melon. $C_3H_6N_6$ can be converted into other derivatives by thermal condensation [24–27]. The synthesized product was subjected to annealing in a vacuum furnace to enhance the GaN crystallinity and eliminate the residual $C_3H_6N_6$. The XRD pattern of annealed nanopowder is indicated in Figure 6. Distinct peaks of GaN were observed after annealing at 850 °C for three hours. Weak graphite and carbon nitride (C_3N_4) peaks were also observed. It was estimated that the synthesized GaN nanopowder was not accurately detected in the XRD pattern due to its low crystallinity and low quantity.

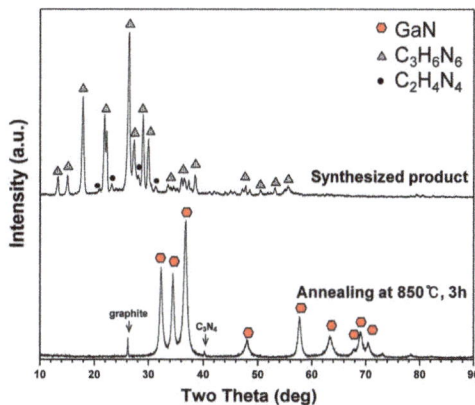

Figure 6. XRD patterns of products synthesized from a $Ga(NO_3)_3 \cdot 32H_2O$ pellet and $C_3H_6N_6$ powder injection with NH_3 gas by the thermal plasma process in Plasma 3.

This result was reliable as confirmed by the Ga–N chemical bonding by XPS results shown in Figure 7. The Ga2p, N1s, and C1s orbital graphs are indicated in Figure 7a,b. In the Ga2p graphs of Figure 7a, Ga–N bonding peaks of $Ga2p_{(1/2)}$ and $Ga2p_{(3/2)}$ were identically observed at 1143.0 eV and 1116.2 eV [26,28–31]. Although peak intensities of $Ga2p_{(1/2)}$ and $Ga2p_{(3/2)}$ in Figure 7a are weak, the two peaks could be accurately observed in the Ga2p graph. This result demonstrated that GaN was definitely synthesized by the thermal plasma, but was not detected in the XRD pattern due to its low crystallinity. The N=C sp2 binding peak is high (398.7 eV) in the N1s graphs of Figure 7. This peak was caused by residual $C_3H_6N_6$ and its derivatives after synthesis using thermal plasma [32]. An N–Ga bonding peak was observed at 397.8 eV. Amino functional groups having ($-NH_x$) were indicated at 400.2 eV by residual $C_3H_6N_6$ and it derivatives. In the C1s graph, C=N and C–C binding peaks were present at 287.5 and 284.3 eV, respectively, which was attributed to the remaining $C_3H_6N_6$. Additionally, the intensity of the C=N bonding peak was higher than that of the C-C bonding peak [32].

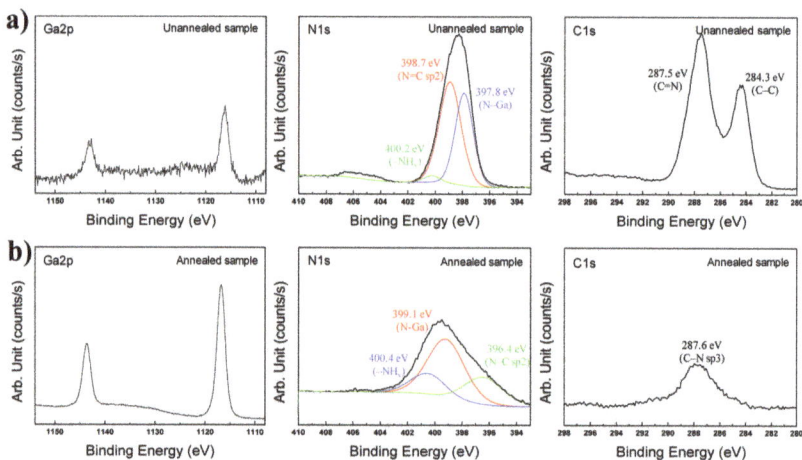

Figure 7. XPS (X-ray photoelectron spectroscopy analysis) of nanopowder synthesized from a $Ga(NO_3)_3 \cdot 32H_2O$ pellet and $C_3H_6N_6$ powder injection with NH_3 gas by the thermal plasma process in Plasma 3; (**a**) before and (**b**) after annealing.

Figure 7b shows XPS analysis results of annealed nanopowder after thermal plasma synthesis in Plasma 3. XRD and TEM analysis revealed that the GaN synthesized by the thermal plasma had enhanced crystallinity after annealing. After annealing at 850 °C under rough vacuum pressure for three hours, the peak intensities of Ga–N bonding were increased to levels which were significantly higher than before the annealing step. Ga–N bonding peaks of $Ga2p_{(1/2)}$ and $Ga2p_{(3/2)}$ were observed at 1143.5 eV and 1116.7 eV, respectively [26,28–31]. However, the N1s graphs all show different peak trends, as is shown in Figure 7a. Following annealing of the product synthesized in Plasma 3, the peaks appeared at 400.4 and 396.4 eV in N1s, which can be observed in Figure 7b. This peak was attributed to nitrogen in the $-NH_x$ and N–C sp2 bond from C_3N_4 generated from the conversion of residual $C_3H_6N_6$ [33,34]. A deconvoluted N–Ga bonding peak was observed at 399.1 eV [27]. This peak was shifted compared to pre-annealing due to the sufficient levels of nitrogen of synthesized GaN by decomposition of residual $C_3H_6N_6$. In the C1s graph of Figure 7b, a C–N sp3 bonding peak was produced at 287.6 eV. C–C bonding peaks are deconvoluted at 289.8 and 284.8 eV [33,34].

The results of TGA analysis of product synthesized under the Plasma 3 operating conditions are indicated in Figure 8. The temperature of the crucible containing the synthesized product powder was increased from 40 °C to 850 °C at 10 °C/min intervals. The atmosphere was filled with a nitrogen gas. A total mass reduction of up to 98% of weight occurred. The "as prepared product" after TGA analysis was considered to consist of GaN and a small amount of C_3N_4. As reported in previous studies [26,33,34], C_3N_4 can be synthesized at high temperatures (>600 °C) by slowly heating $C_3H_6N_6$ alone. The mass reduction curve had a high gradient from about 300 °C to 650 °C. The residual $C_3H_6N_6$ and derivatives undergo thermal degradation and decomposition at temperatures in excess of 300 °C. Based on the mass maintenance at above 620 °C, the temperature for annealing to enhance the crystallinity of synthesized GaN nanopowder was set at 850 °C. Since the melting point of GaN is above 2500 °C, synthesized GaN nanopowder was not melted or vaporized by the annealing.

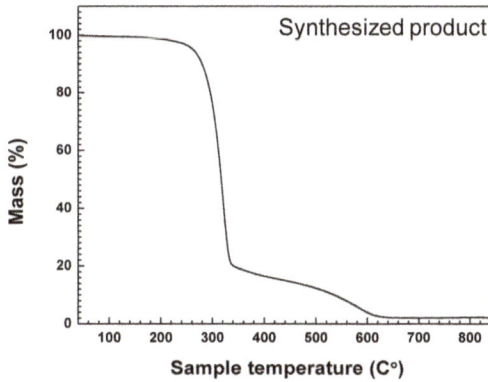

Figure 8. TGA analysis of product synthesized from a $Ga(NO_3)_3 \cdot 32H_2O$ pellet and $C_3H_6N_6$ powder injection with NH_3 gas by the thermal plasma process in Plasma 3.

Figure 9a,b shows FE-TEM images of the synthesized product before and after annealing in Plasma 3 conditions. Small particles under 100 nm and spherical large particles of about 200 nm were mixed in Figure 9a. After annealing, the large spherical particles were no longer present, and only nano-sized round particles which had an average particle size of 21.63 nm remained, as is shown in Figure 9b. It is believed that the large spherical particles are the residual $C_3H_6N_6$ precursor. Figure 9c shows TEM-EDS results before and after annealing in Plasma 3. The Ga element was detected as about 2.5% of weight before annealing. It was revealed that melamine powder remained in micro-sized particles. The content of Ga elements was increased dramatically from 2.45% to 71.6% of weight after annealing. This analysis demonstrates that the purification was completed and residual melamine powder was eliminated by the heat treatment of the vacuum furnace. As shown by the XRD patterns in Figure 6b, a small amount of carbon and carbon nitride was generated after annealing because the treatment was done under vacuum conditions. The vaporized melamine converted as a byproduct with gallium nitride nanoparticles.

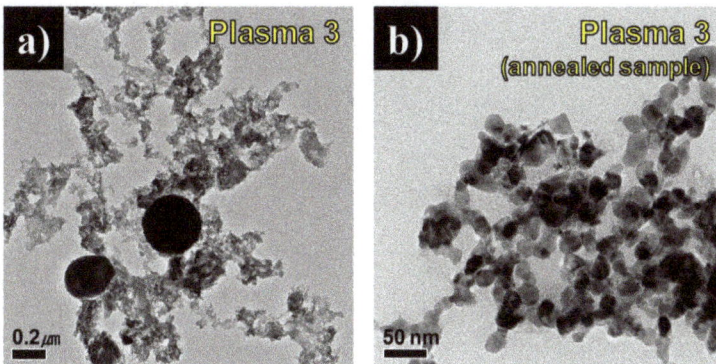

Figure 9. *Cont.*

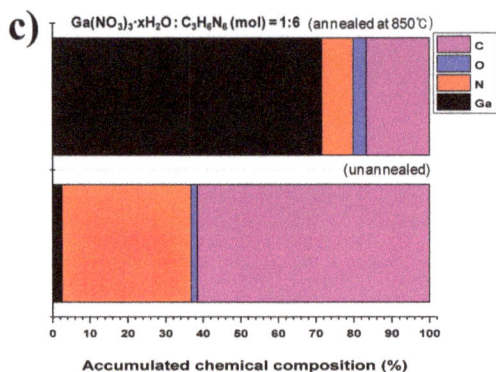

Figure 9. FE-TEM (field-emission scanning electron microscopy) and EDS (energy dispersive spectroscopy) results of product synthesized from a $Ga(NO_3)_3 \cdot 32H_2O$ pellet and $C_3H_6N_6$ powder injection with NH_3 gas by the thermal plasma process in Plasma 3; FE-TEM images of (**a**) before and (**b**) after annealing; (**c**) TEM-EDS results.

FE-SEM images and size distribution of synthesized GaN nanopowder are indicated in Figure 10a,b. The large residual $C_3H_6N_6$ particles were removed, but the uniform GaN nanoparticles remained, as is shown in Figure 10a. The particle size distribution was arranged by measurement of each particle in the FE-SEM frame, as is seen in Figure 10b. The size distribution of synthesized nanoparticles was analyzed by FE-SEM images. Hundreds of nanoparticles were measured with particles sizes varying from 10 to 60 nm. The mean particle size was 29.8 nm. The uniform-sized particle GaN nanopowder was synthesized by thermal plasma and an additional annealing process.

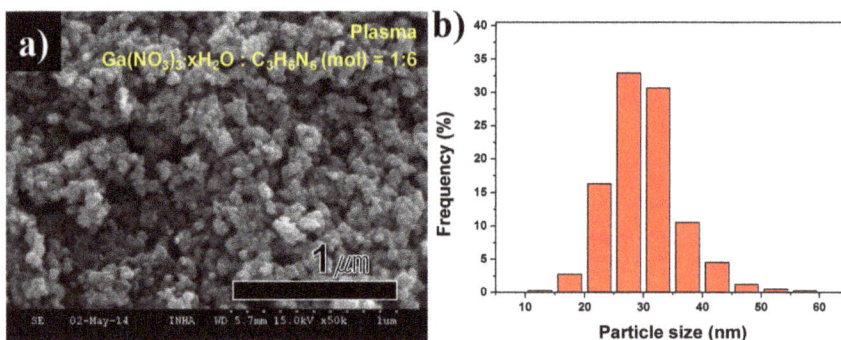

Figure 10. (**a**) FE-SEM image and (**b**) size distribution of synthesized GaN nanopowder after annealing in Plasma 3.

To reduce residual $C_3H_6N_6$ present in the product generated by the thermal plasma, the ratio of $Ga(NO_3)_3 \cdot 32H_2O$ and $C_3H_6N_6$ was decreased to 1:3 under the operating conditions used to generate Plasma 4. XRD patterns of product synthesized under these conditions are shown in Figure 11a. The upper pattern is for synthesized nanopowder by thermal plasma while the bottom pattern is for annealed nanopowder at 850 °C for three hours. In the upper XRD pattern, the main peak of $C_3H_6N_6$ was observed at a weak 26°, while main peaks of GaN were found at 32°, 34°, and 38°. However, Ga_2O_3 peaks appeared after annealing in the vacuum furnace at 850 °C for three hours.

These findings demonstrate that the oxygen capture reaction of $Ga(NO_3)_3 \cdot 32H_2O$ by $C_3H_6N_6$ could have negative effects as the amount of $C_3H_6N_6$ decreases.

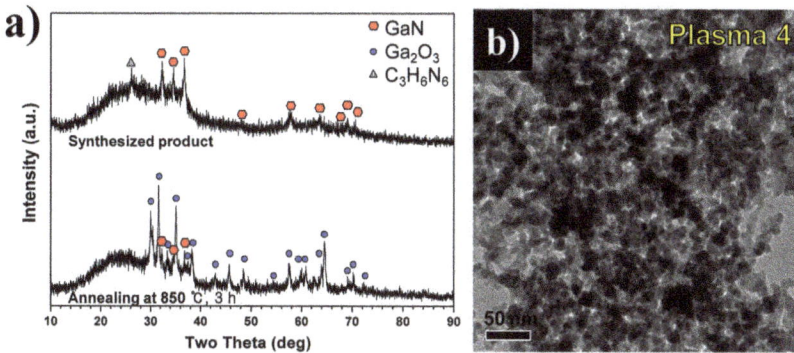

Figure 11. (a) XRD patterns before and after annealing and (b) FE-TEM image of unannealed product from a $Ga(NO_3)_3 \cdot 32H_2O$ pellet and $C_3H_6N_6$ powder injection with NH_3 gas by the thermal plasma process in Plasma 4.

Figure 11b shows FE-TEM images of synthesized nanopowder by thermal plasma, when the molar ratio of $Ga(NO_3)_3 \cdot 32H_2O$ and $C_3H_6N_6$ of 1:3 is adopted in Plasma 4 conditions. The product synthesized in Plasma 4 primarily consisted of fine particles which had an average particle size of 9.25 nm. In the EDS result in Table 2, the content of the Ga element is higher, about 63% of weight, than that of the synthesized nanopowder in Plasma 3 without additional annealing. However, the oxygen content is high at 23% of weight due to incomplete nitridation and generated gallium oxide. The degree of Ga oxidation increased as the molar ratio of injected $C_3H_6N_6$ decreased. Therefore, injection of excessive $C_3H_6N_6$ is effective for preventing the oxidization of Ga. Moreover, $C_3H_6N_6$ aids in the complete synthesizing of GaN nanopowder. However, annealing is required to eliminate the remaining $C_3H_6N_6$.

Table 2. Elemental composition of product synthesized from a $Ga(NO_3)_3 \cdot 32H_2O$ pellet and $C_3H_6N_6$ powder injection with NH_3 gas by the thermal plasma process in Plasma 4 (all results in % of weight).

Spectrum	C	N	O	Ga	Total
Spectrum 1	8.70	3.51	23.04	64.76	100.00
Spectrum 2	12.60	4.78	23.34	59.28	100.00
Spectrum 3	9.05	1.73	23.40	65.82	100.00
Mean	10.11	3.34	23.26	63.28	100.00

$C_3H_6N_6$ was used as a reductant to prevent oxidation of gallium. However, it also consists of an abundant source of nitrogen. Therefore, a nitridation reaction using only $C_3H_6N_6$ powder, without the addition of NH_3 is needed to confirm the complete synthesis of GaN nanopowder. In previous experiments, NH_3 gas was injected into the $Ga(NO_3)_3 \cdot 32H_2O$ pellet as a nitridation source. A $Ga(NO_3)_3 \cdot 32H_2O$ pellet was reacted with only $C_3H_6N_6$ powder without NH_3 gas in Plasma 5 conditions. $C_3H_6N_6$ powder was injected through two nozzles of an anode electrode as in Plasma 3. The molar ratio of $Ga(NO_3)_3 \cdot 32H_2O$ and $C_3H_6N_6$ was set at 1:6. The XRD pattern of synthesized nanopowder by reacting $Ga(NO_3)_3 \cdot 32H_2O$ and $C_3H_6N_6$ without NH_3 gas is indicated in Figure 12a. The intensities of all peaks were weak. The observed peaks corresponded with gallium oxide and melamine. When the NH_3 gas was injected into the reactor at the same molar ratio of $Ga(NO_3)_3 \cdot 32H_2O$ and $C_3H_6N_6$ in Plasma 3, only residual $C_3H_6N_6$ peaks were detected in the XRD pattern in Figure 6.

Low crystallinity GaN was synthesized. However, Ga_2O_3 peaks were detected with residual $C_3H_6N_6$ in Plasma 5 of Figure 12a. It was revealed that the nitridation reaction did not occur with the nitrogen element of $C_3H_6N_6$ powder. Figure 12b shows XPS analysis results of synthesized nanopowder in Plasma 5. Ga-N bonding peaks as $Ga2p_{(1/2)}$ and $Ga2p_{(3/2)}$ were identically observed at 1143.0 eV and 1116.2 eV. However, two peaks were observed at 1145.9 eV and 1119.3 eV. Analysis of these peaks confirmed Ga–O chemical bonding. An increase of binding energy from Ga2p electrons was observed due to Ga–O chemical bonding. The nitrogen element has a less electronegative property than the oxygen element. Therefore, the binding energy of Ga–O chemical bonding is higher than that of Ga–N chemical bonding [30].

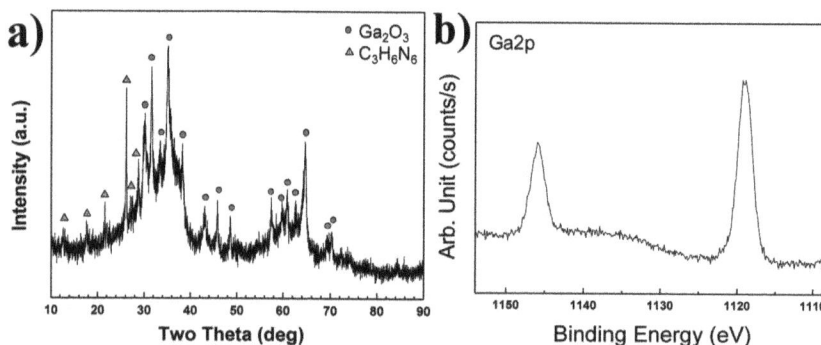

Figure 12. (**a**) XRD patterns and (**b**) XPS analysis of products synthesized from a $Ga(NO_3)_3 \cdot 32H_2O$ pellet and $C_3H_6N_6$ powder injection without NH_3 gas by the thermal plasma process in Plasma 5.

$C_3H_6N_6$ was decomposed and converted to cyanides, hydrocarbons, and carbon in the high temperature of the thermal plasma jet. Among these, only cyanides have nitrogen elements. However, the nitridation reaction by cyanide molecules is an inferior chemical reaction compared with oxidation. It was examined using the thermodynamic equilibrium calculation in Figure 4c.

Overall, these findings indicate that GaN nanopowder was synthesized from $Ga(NO_3)_3 \cdot 32H_2O$ and $C_3H_6N_6$ by thermal plasma, despite low crystallinity. Its crystallinity can be enhanced by annealing in a vacuum furnace. Moreover, if the annealing can be completed at atmospheric pressure in an inert atmosphere, not by rough vacuum, production of carbon dioxide or carbon nitride can be controlled.

4. Conclusions

GaN nanopowder was synthesized by a thermal plasma process. $Ga(NO_3)_3 \cdot xH_2O$ was used as the raw material. At first, it was discovered that $Ga(NO_3)_3 \cdot xH_2O$ was not nitrided by a solely conventional NH_3 nitridation source and that it is converted into Ga_2O_3. It required a reductant to prevent oxidation to Ga_2O_3 instead of GaN. Therefore, $C_3H_6N_6$ powder was injected into the high temperature region of the thermal plasma jet through an anode electrode. The molar ratio of injected $Ga(NO_3)_3 \cdot xH_2O$ and $C_3H_6N_6$ was controlled at 1:6 and 1:3. GaN nanopowder with low crystallinity and residual $C_3H_6N_6$ was synthesized by the thermal plasma process. Crystallinity of the synthesized GaN nanopowder was further enhanced after annealing at 850 °C for three hours in a vacuum furnace. The size of synthesized GaN nanopowder is distributed from 10 to 60 nm, and mean particle size is calculated to be 29.8 nm. In order to confirm the nitridation of $Ga(NO_3)_3 \cdot xH_2O$ by $C_3H_6N_6$ reductant, the $Ga(NO_3)_3 \cdot xH_2O$ was reacted with $C_3H_6N_6$ without NH_3 gas, and $Ga(NO_3)_3 \cdot xH_2O$ precursor was oxidized to Ga_2O_3. Therefore, GaN nanopowder was successfully synthesized from $Ga(NO_3)_3 \cdot 32H_2O$ and $C_3H_6N_6$ powders with NH_3 gas by a thermal plasma process. Furthermore, the additional annealing step was required to enhance its crystallinity. It was speculated that the production of carbon dioxide or carbon nitride during the annealing step could be controlled as anneal in an inert atmosphere. The synthesis of GaN crystalline nanopowder is difficult to achieve using conventional production methods. This research demonstrates the potential to synthesize GaN nanopowder through a thermal plasma process, from raw materials comprising abundant oxygen elements.

Acknowledgments: Acknowledgments: This research was supported by the World Class 300 Project (10043264, Development of the electrode materials for high efficiency (21%) and low-cost c-Si solar cells) funded by the Ministry of Knowledge Economy of the Republic of Korea, and by the Basic Science Research Program through the National Research Foundation of Korea (NRF) funded by the Ministry of Education of the Republic of Korea (No. NRF-2010-0020077).

Author Contributions: Author Contributions: T.-H.K. and S.C. conceived and designed the experiments; T.-H.K. performed the experiments; T.-H.K. and D.-W.P. analyzed the data; All the three authors wrote the paper.

Conflicts of Interest: Conflicts of Interest: The authors declare no conflict of interest.

1. Morkoc, H.; Mohammad, S.N. High-luminosity blue and blue-green gallium nitride light-emitting diodes. *Science* **1995**, *267*, 51–55. [CrossRef] [PubMed]
2. Balkas, C.M.; Davis, R.F. Synthesis routes and characterization of high-purity, single-phase gallium nitride powders. *J. Am. Ceram. Soc.* **1996**, *79*, 2309–2312. [CrossRef]
3. Han, W.; Fan, S.; Li, Q.; Hu, Y. Synthesis of gallium nitride nanorods through a carbon nanotube-confined reaction. *Science* **1997**, *277*, 1287–1289. [CrossRef]
4. Cheng, G.S.; Zhang, L.D.; Zhu, Y.; Fei, G.T.; Li, L. Large-scale synthesis of single crystalline gallium nitride nanowires. *Appl. Phys. Lett.* **1999**, *75*, 2455–2457. [CrossRef]
5. Mohammad, S.N.; Salvador, A.A.; Morkoc, H. Emerging gallium nitride based devices. *Proc. IEEE* **1995**, *83*, 1306–1355. [CrossRef]
6. Kim, H.M.; Kang, T.W.; Chung, K.S. Nanoscale ultraviolet-light-emitting diodes using wide-bandgap gallium nitride nanorods. *Adv. Mater.* **2003**, *15*, 567–569. [CrossRef]
7. Millet, P.; Calka, A.; Williams, J.S.; Vantenaar, G.J.H. Formation of gallium nitride by a novel hot mechanical alloying process. *Appl. Phys. Lett.* **1993**, *63*. [CrossRef]
8. Wu, H.; Hunting, J.; Uheda, K.; Lepak, L.; Konkapaka, P.; DiSalvo, F.J.; Spencer, M.G. Rapid synthesis of gallium nitride powder. *J. Cryst. Growth* **2005**, *279*, 303–310. [CrossRef]
9. Lorenz, M.R.; Binkowski, B.B. Preparation, stability, and luminescence of gallium nitride. *J. Electrochem. Soc.* **1962**, *109*, 24–26. [CrossRef]
10. Wood, G.L.; Pruss, E.A.; Paine, R.T. Aerosol-assisted vapor phase synthesis of gallium nitride powder. *Chem. Mater.* **2001**, *13*, 12–14. [CrossRef]
11. Li, H.D.; Yang, H.B.; Yu, S.; Zou, G.T.; Li, Y.D.; Liu, S.Y.; Yang, S.R. Synthesis of ultrafine gallium nitride powder by the direct current arc plasma method. *Appl. Phys. Lett.* **1996**, *69*, 1285–1287. [CrossRef]
12. Li, H.D.; Yang, H.B.; Zou, G.T.; Yu, S.; Lu, J.S.; Qu, S.C.; Wu, Y. Formation and photoluminescence spectrum of w-GaN powder. *J. Cryst. Growth* **1997**, *171*, 307–310. [CrossRef]
13. Chen, C.C.; Yeh, C.C.; Chen, C.H.; Yu, M.Y.; Liu, H.L.; Wu, J.J.; Chen, K.H.; Chen, L.C.; Peng, J.Y.; Chen, Y.F. Catalytic growth and characterization of gallium nitride nanowires. *J. Am. Chem. Soc.* **2001**, *123*, 2791–2798. [CrossRef] [PubMed]
14. Jung, W.S. Reaction intermediate(s) in the conversion of β-gallium oxide to gallium nitride under a flow of ammonia. *Mater. Lett.* **2002**, *57*, 110–114. [CrossRef]
15. Jung, W.S.; Min, B.K. Synthesis of gallium nitride powders and nanowires from gallium oxyhydroxide under a flow of ammonia. *Mater. Lett.* **2004**, *58*, 3058–3062. [CrossRef]
16. Jung, W.S. Synthesis and characterization of GaN powder by the cyanonitridation of gallium oxide powder. *Ceram. Int.* **2012**, *38*, 5741–5746. [CrossRef]
17. Melnikov, P.; Nascimento, V.A.; Zanoni Consolo, L.Z. Thermal decomposition of gallium nitrate hydrate and modeling of thermolysis products. *J. Therm. Anal. Calorim.* **2012**, *107*, 1117–1121. [CrossRef]
18. Ogi, T.; Kaihatsu, Y.; Iskandar, F.; Tanabe, E.; Okuyama, K. Synthesis of nanocrystalline GaN from Ga_2O_3 nanoparticles derived from salt-assisted spray pyrolysis. *Adv. Powder Technol.* **2009**, *20*, 29–34. [CrossRef]
19. Di Lello, B.C.; Moura, F.J.; Solorzano, I.G. Synthesis and characterization of GaN using gas-solid reactions. *Mater. Sci. Eng. B* **2002**, *93*, 219–223. [CrossRef]

20. Jung, W.S. Preparation of gallium nitride powders and nanowires from a gallium(III) nitrate salt in flowing ammonia. *Bull. Korean Chem. Soc.* **2004**, *25*, 51–54.

21. Zhao, H.; Lei, M.; Chen, X.; Tang, W. Facile route to metal nitrides through melamine and metal oxides. *J. Mater. Chem.* **2006**, *16*, 4407–4412. [CrossRef]

22. Jung, W.S. Use of melamine in the nitridation of aluminium oxide to aluminium nitride. *J. Ceram. Soc. Jpn.* **2012**, *119*, 968–971. [CrossRef]

23. Berbenni, V.; Milanese, C.; Bruni, G.; Marini, A. Thermal decomposition of gallium nitrate hydrate $Ga(NO_3)_3 \cdot 32H_2O$. *J. Therm. Anal. Calorim.* **2005**, *82*, 401–407. [CrossRef]

24. Ono, S.; Funato, T.; Inoue, Y.; Munechika, T.; Yoshimura, T.; Morita, H.; Rengakuji, S.I.; Shimasaki, C. Determination of melamine derivatives, melame, meleme, ammeline and ammelide by high-performance cation-exchange chromatography. *J. Chromatogr. A* **1998**, *815*, 197–204. [CrossRef]

25. Costa, L.; Camino, G. Thermal behavior of melamine. *J. Therm. Anal.* **1988**, *34*, 423–429. [CrossRef]

26. Yan, H.; Chen, Y.; Xu, S. Synthesis of graphitic carbon nitride by directly heating sulfuric acid treated melamine for enhanced photocatalytic H_2 production from water under visible light. *Int. J. Hydrog. Energy* **2012**, *37*, 125–133. [CrossRef]

27. Xiao, H.D.; Ma, H.L.; Xue, C.S.; Hu, W.R.; Ma, J.; Zong, F.J.; Zhang, X.J.; Ji, F. Synthesis and structural properties of GaN particles from GaO_2H powders. *Diam. Relat. Mater.* **2005**, *14*, 1730–1734. [CrossRef]

28. Wolter, S.D.; Luther, B.P.; Waltemyer, D.L.; Onneby, C.; Mohney, S.E.; Molnar, F.J. X-ray photoelectron spectroscopy and x-ray diffraction study of the thermal oxide on gallium nitride. *Appl. Phys. Lett.* **1997**, *70*, 2156–2158. [CrossRef]

29. Yang, Y.; Ma, H.; Hao, X.; Ma, J.; Xue, C.; Zhuang, H. Preparation and properties of GaN films on Si(111) substrates. *Sci. China Ser. G* **2003**, *46*, 173–177. [CrossRef]

30. Pal, S.; Mahapatra, R.; Ray, S.K.; Chakraborty, B.R.; Shivaprasad, S.M.; Lahiri, S.K.; Bose, D.N. Microwave plasma oxidation of gallium nitride. *Thin Solid Film.* **2003**, *425*, 20–23. [CrossRef]

31. Xiao, H.D.; Ma, H.L.; Xue, C.S.; Zhuang, H.Z.; Ma, J.; Zong, F.J.; Zhang, X.J. Synthesis and structural properties of beta-gallium oxide particles from gallium nitride powder. *Mater. Chem. Phys.* **2007**, *101*, 99–102. [CrossRef]

32. Feng, D.; Zhou, Z.; Bo, M. An investigation of the thermal degradation of melamine phosphonate by XPS and thermal analysis techniques. *Polym. Degrad. Stable* **1995**, *50*, 65–70. [CrossRef]

33. Yao, L.D.; Li, F.Y.; Li, J.X.; Jin, C.Q.; Yu, R.C. Study of the products of melamine ($C_3H_6N_6$) treated at high pressure and high temperature. *Phys. Status Solidi (a)* **2005**, *202*, 2679–2685. [CrossRef]

34. Li, X.; Zhang, J.; Shen, L.; Ma, Y.; Lei, W.; Cui, Q.; Zou, G. Preparation and characterization of graphite carbon nitride through pyrolysis of melamine. *Appl. Phys. A* **2009**, *94*, 387–392. [CrossRef]

nanomaterials

MDPI

Article

Resistive Switching of Plasma–Treated Zinc Oxide Nanowires for Resistive Random Access Memory

Yunfeng Lai [1,2,*], Wenbiao Qiu [1], Zecun Zeng [1], Shuying Cheng [1], Jinling Yu [1] and Qiao Zheng [1]

[1] School of Physics and Information Engineering, Fuzhou University, Fuzhou 350108, China; albert_29@163.com (W.Q.); zecunzeng@163.com (Z.Z.); sycheng@fzu.edu.cn (S.C.); jlyu@semi.ac.cn (J.Y.); 2004_zhengqiao@163.com (Q.Z.)
[2] Jiangsu Collaborative Innovation Center of Photovoltaic Science and Engineering, Changzhou University, Changzhou 213164, China
* Correspondence: yunfeng.lai@fzu.edu.cn; Tel.: +86-591-22866342

Academic Editors: Krasimir Vasilev and Melanie Ramiasa
Received: 30 November 2015; Accepted: 8 January 2016; Published: 13 January 2016

Abstract: ZnO nanowires (NWs) were grown on Si(100) substrates at 975 °C by a vapor-liquid-solid method with ~2 nm and ~4 nm gold thin films as catalysts, followed by an argon plasma treatment for the as-grown ZnO NWs. A single ZnO NW–based memory cell with a Ti/ZnO/Ti structure was then fabricated to investigate the effects of plasma treatment on the resistive switching. The plasma treatment improves the homogeneity and reproducibility of the resistive switching of the ZnO NWs, and it also reduces the switching (set and reset) voltages with less fluctuations, which would be associated with the increased density of oxygen vacancies to facilitate the resistive switching as well as to average out the stochastic movement of individual oxygen vacancies. Additionally, a single ZnO NW–based memory cell with self-rectification could also be obtained, if the inhomogeneous plasma treatment is applied to the two Ti/ZnO contacts. The plasma-induced oxygen vacancy disabling the rectification capability at one of the Ti/ZnO contacts is believed to be responsible for the self-rectification in the memory cell.

Keywords: resistive switching; plasma treatment; ZnO nanowires; self-rectification

---

1. Introduction

Resistive random access memory (RRAM) is capable of reversible switching between high resistance state (HRS) and low resistance state (LRS) under suitable electrical stress, which has been confirmed to be closely associated with defects in the memory [1–3]. The significance of controlling defects in the memory cells would be more pronounced in the nano-era, since the memory cells shrink their physical dimension for high density integration [4–6]. As for the storage medium of the RRAMs, zinc oxide (ZnO), a wide band gap semiconductor with abundant intrinsic defects, has recently attracted considerable attention for its potential applications in transparent RRAMs due to its excellent resistive switching [7–9]. Understanding and controlling the defects in the ZnO nanowire (NW)–based RRAM are thus worthwhile [6,10–12].

In order to further improve storage performance such as the retention, stability and homogeneity of switching properties, several schemes have been attempted to control defects in the storage medium and on the interface [13–19]. Doping certain elements or embedding nano-particles into the storage medium seems to be a feasible way to rearrange the defects and the storage performance could thus be improved [13–15]. An alternative way is to change interfacial defects by selecting suitable electrodes or by stacking a different storage medium to form a hetero-interface [16–19]. Plasma treatment is a widely accepted technique for surface modification due to its physical effects as well as chemical ones on the treated materials. It also exhibits low cost, high efficiency, high control accuracy and good

![nanomaterials logo]

nanomaterials

MDPI

Article

Dense Plasma Focus-Based Nanofabrication of III–V Semiconductors: Unique Features and Recent Advances

Onkar Mangla [1,2], Savita Roy [3,*] and Kostya (Ken) Ostrikov [4,5,6]

1   Department of Physics and Astrophysics, University of Delhi, Delhi 110007, India; onkarmangla@gmail.com
2   Physics Department, Hindu College, University of Delhi, Delhi 110007, India
3   Physics Department, Daulat Ram College, University of Delhi, Delhi 110007, India
4   School of Chemistry, Physics and Mechanical Engineering, Queensland University of Technology (QUT), Brisbane 4000, Australia; kostya.ostrikov@csiro.au
5   Plasma Nanoscience Laboratories, Commonwealth Scientific and Industrial Research Organisation, P.O. Box 218, Lindfield 2070, Australia
6   Plasma Nanoscience, School of Physics, The University of Sydney, Sydney 2006, Australia
*   Correspondence: savitaroy64@gmail.com; Tel.: +91-9810629598

Academic Editors: Krasimir Vasilev and Melanie Ramiasa
Received: 12 October 2015; Accepted: 17 December 2015; Published: 29 December 2015

Abstract: The hot and dense plasma formed in modified dense plasma focus (DPF) device has been used worldwide for the nanofabrication of several materials. In this paper, we summarize the fabrication of III–V semiconductor nanostructures using the high fluence material ions produced by hot, dense and extremely non-equilibrium plasma generated in a modified DPF device. In addition, we present the recent results on the fabrication of porous nano-gallium arsenide (GaAs). The details of morphological, structural and optical properties of the fabricated nano-GaAs are provided. The effect of rapid thermal annealing on the above properties of porous nano-GaAs is studied. The study reveals that it is possible to tailor the size of pores with annealing temperature. The optical properties of these porous nano-GaAs also confirm the possibility to tailor the pore sizes upon annealing. Possible applications of the fabricated and subsequently annealed porous nano-GaAs in transmission-type photo-cathodes and visible optoelectronic devices are discussed. These results suggest that the modified DPF is an effective tool for nanofabrication of continuous and porous III–V semiconductor nanomaterials. Further opportunities for using the modified DPF device for the fabrication of novel nanostructures are discussed as well.

Keywords: III–V semiconductors; nanofabrication; dense plasma focus; rapid thermal annealing; photoluminescence; transmittance

1. Introduction

Nanostructured materials have gained strong recent interest due to their superior properties as compared to their thin films and bulk counter parts. These properties of nanostructures result in several advanced applications which are not possible using the same materials in their bulk and thin film forms. However, these properties and applications of nanostructures depend upon the method of nanofabrication.

There are several chemical and physical methods that have successfully been used for fabrication of nanostructures of different materials. These chemical and physical methods have several disadvantages like slow deposition rate, contaminants from precursors and catalysts, substrate heating or biasing during deposition, post annealing of deposited material, *etc*. The plasma based methods are often found to be superior to the similar chemical and physical methods.

There are several plasma based nanofabrication methods, such as arc discharge, direct current, radio frequency and magnetron sputtering, pulsed laser deposition, and modified dense plasma focus (DPF) devices, which have been successfully used for the fabrication of a broad range of nanomaterials. Most of these plasma-based methods have certain disadvantages. For example, some of them require substrate heating or biasing, ultrahigh vacuum, or show poor yield, abundant pinholes, some other defects, *etc.* Hot, dense, and extremely non-equilibrium plasmas in modified DPF device overcome most of the above disadvantages and also reduces time for nanofabrication.

The modified DPF device with suitable modification has been used worldwide for the fabrication of nanoparticles and nanostructures of different materials [1–21]. In most of these studies, the modification of anode is almost similar but the arrangement for placing the substrate varies. The modified DPF device has been established as a promising tool for nanofabrication by the plasma research group at the University of Delhi [7–21]. In particular, several nanostructures of metals [8–13], semiconductors [14–18] and insulators [19–21] have been produced using the modified DPF device. Apart from this the nanostructures of materials, such as TiC [3], hydroxyapatite (HA) [4], Al/a-C [5], WN_2 [6], *etc.*, have been fabricated worldwide with the modified DPF device.

The semiconductor materials when fabricated at nanoscale show drastic change in their optical and electronic properties. These changes are mainly due to the quantum confinement effect in the semiconductor nanostructures. There are several semiconductor materials that belong to different groups of the periodic table. The III–V semiconductors are preferred over other semiconductors for microelectronic applications owing to their wide and direct band gap, high electron mobility, low thermal noise, as well as low power consumption.

These properties of III–V semiconductors results in fabrication of devices having wide applications. For example, direct band gap of III–V semiconductors is easily tunable at nanoscale which gives rise to potential applications of the nanostructures in optoelectronic devices operated in broad spectral ranges.

The above properties vary among different nanofabrication methods. The III–V semiconductor nanostructures of materials like gallium nitride (GaN) and gallium arsenide (GaAs) having different morphologies such as nanoparticles [22,23], nanorods [24,25], nanocrystals [26,27], nanopillars [28,29], nanowires [30,31], nanotubes [32,33], nanobelts [34,35], nanoneedles [36,37], *etc.*, have been fabricated by methods such as sol-gel [22], hydride vapor phase epitaxy (HVPE) [38], arc plasma [26], RF sputtering [23,39], inductively coupled plasma (ICP) [28], thermal evaporation [40], electrochemical [41], molecular beam epitaxy (MBE) [42], pulsed laser ablation [43], pulsed electron deposition [44], pulsed laser deposition (PLD) [45], laser-assisted catalytic growth [46], and modified DPF devices [16–18].

Most of these methods have limitations, such as slow rates of deposition, contaminants arising out of precursors and catalysts, large power consumption during the heating, and biasing of substrate, as well as the need for ultra-high vacuum and post-deposition annealing. These limitations have been overcome using the modified DPF device for nanoscale synthesis.

The modified DPF device has been successfully used for the fabrication of nanostructures of III–V semiconductors, such as GaN [16] and GaAs [17,18], without any substrate heating/biasing and annealing of deposited material. In particular, GaAs has potential applications in photovoltaic [44], electronic and optoelectronic [17,47,48] devices. Apart from the GaAs nanostructures, the porous GaAs can also be used as a substrate for the growth of other semiconducting materials, such as gallium nitride [49,50]. Moreover, this material produces emission in the visible range which can be used for sensors, light emitting and plasmonic devices. The porous GaAs can also increase the efficiency of the photoelectric and photovoltaic devices, such as solar cells, due to reduced optical losses and other interesting properties.

The formation of porous GaAs has been reported in the literature mainly from the chemical based methods which involve electrolysis and etching to form porous GaAs [49–54]. To the best of our knowledge, the formation of porous GaAs using physical or plasma based methods without any etching or electrolysis has not been reported yet. In addition, the fabrication of any porous nanostructures using the modified DPF device has not yet been reported previously.

In this paper, we highlight the applications of modified DPF device for nanoscale synthesis and processing of GaAs on different substrates under different deposition conditions. In particular, we present for the first time fabrication of porous nano-GaAs using the modified DPF device with two shots on glass substrates placed at 4.0 cm distance. Subsequently, we present the characterization results obtained for the as-grown and rapid thermally annealed porous nano-GaAs.

2. Results and Discussion

GaAs nanostructures were deposited on silicon and glass substrates using high fluence ions in the modified DPF device. The morphology of GaAs nanostructures deposited on silicon substrate placed at a distance of 4.0 cm indicates the formation of triangular nanostructures mapped by the atomic force microscopy (AFM), as shown in Figure 1. The dimensions of nanostructures produced using 1, 2, and 3 focused DPF shots are found to be in the range of 10–30 nm, 20–50 nm and 40–70 nm, respectively. Therefore, the dimensions of nanostructures increased with the increase in the number of DPF shots.

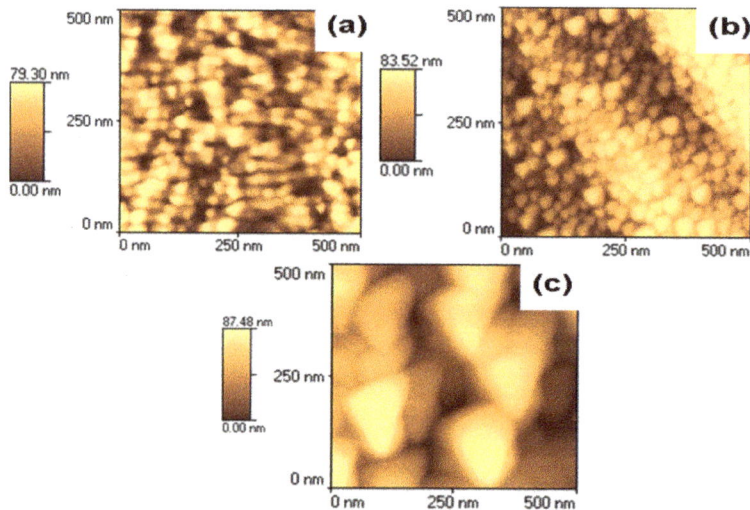

Figure 1. Atomic force microscopy (AFM) images of gallium arsenide (GaAs) nanostructures for (**a**) one, (**b**) two, (**c**) three shots on a silicon substrate placed at a distance of 4.0 cm.

This may be understood as follows. The material is deposited on the substrate randomly in each shot. In the first shot, the material will deposit on the substrate randomly whereas in the second shot the material is either deposited on as-fabricated nanostructures or on the substrate areas having no deposited material on it. In the third shot the material is deposited by the process similar to that in the second shot, which will results into increased dimensions of the nanostructures.

Figure 1 also shows that the surface density of the nanostructures decreases with increasing the number of shots, which is due to the larger sizes and possible agglomeration of the nanostructures. On the other hand, the morphology of GaAs nanostructures on silicon substrate at 5.0 cm distance shows the formation of nanostructures of arbitrary shape, nanodots and agglomerated nanodots in one, two and three shots with the dimensions of nanostructures in the range of 10–30 nm, 20–40 nm and 40–60 nm, respectively.

The nanostructures produced on a silicon substrate placed at 4.0 cm and 5.0 cm were further studied for their structural, optical and electrical properties. These properties of GaAs nanodots, deposited with two shots on silicon placed at a 5.0 cm distance, have been reported previously [17]. These results are almost similar for one and three shots cases at a 5.0 cm distance. On the other hand, the PL spectra for triangular GaAs nanostructures shows peaks related to band edge emission and surface recombinations. These peaks are similar to the peaks obtained at a 5.0 cm distance [17]. A typical photoluminescence (PL) spectrum of triangular nanostructures is shown in Figure 2. The PL peak due to surface recombinations is split at about 1.6 eV to show two peaks at 1.59 eV and 1.61 eV. These two peaks are due to surface recombination processes and are in good agreement with the peaks reported by Nayak *et al.* [41].

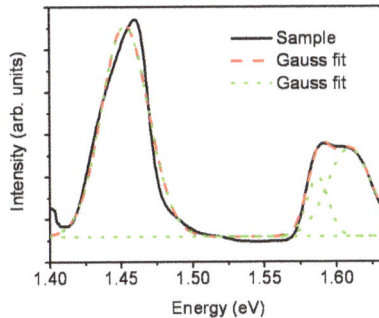

Figure 2. Typical photoluminescence (PL) spectrum of GaAs nanostructures deposited on a silicon substrate placed at a distance of 4.0 cm.

The absorption spectrum of triangular GaAs nanostructures shown in Figure 3a is similar to the spectrum of the nanostructures deposited at a 5.0 cm distance [17]. However, the absorption spectra at 4.0 cm do not possess any excitonic feature as observed at a 5.0 cm distance. Moreover, the band-gap values obtained from Tauc plot at 4.0 cm distance (~3.06 eV) are smaller compared to the 5.0 cm distance (~4.62 eV) as shown in Figure 3b,c, respectively. The electrical properties of the nanostructures deposited on silicon at both 4.0 cm and 5.0 cm distances are identical giving the resistivity values in the range of 7–8 $\Omega \cdot$ cm [17].

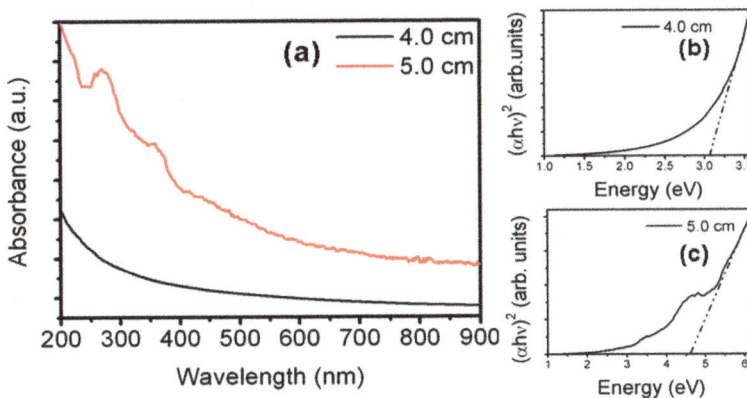

Figure 3. (a) Absorption spectra of GaAs nanostructures deposited on silicon substrate placed at a distance of 4.0 cm and 5.0 cm. Tauc plot showing band gap for (b) 4.0 cm, (c) 5.0 cm distances.

The GaAs nanodots were also fabricated on glass substrate with two shots at a 5.0 cm distance. The morphological and optical studies of these GaAs nanodots have been reported [18]. The nanodots of 6, 10, and 13 nm sizes were smaller than the Bohr exciton diameter (30 nm) of GaAs. This has resulted in the quantum confinement effect and a clear shift of the band-gap from 1.43 eV for bulk GaAs to 2.88, 2.60, and 2.40 eV for 6, 10, and 13 nm, respectively [18].

The effect of effective mass ratio of electrons and holes has been discussed and it was reported that the deviation between the experimental and theoretical band gap values is not due to approximated spherical shape but it is due to the change in effective mass ratio at nanoscale [18]. On the other hand, if the glass substrates are placed at a distance of 4.0 cm from the top of the anode, then porous nano-GaAs is formed in two shots. The porosity of the nano-GaAs decreases upon rapid thermal annealing of as-deposited (sample A) sample at temperatures of 100 °C (sample B), 200 °C (sample C) and 300 °C (sample D) for 280 ms. The formation of porous nano-GaAs using modified DPF device has been observed for first time and some of the original results on its characterization are presented below. In addition, it is possible to optimize the temperature of annealing for the given pores size but in the present study we investigated the effect of annealing on pores size.

Figures 4a, 5a and 6a show the scanning electron microscopy (SEM) images of sample A, B and C, respectively, which indicates the formation of porous GaAs with uniformly distributed pores. The pore sizes have been found for samples A, B, and C from cross-sectional SEM images in Figures 4b, 5b and 6b, respectively. The cross-sectional SEM images show pores that extend from the surface into the sample. The cross-sectional image of large portion of the sample A, showing upper and lower portions is shown in Figure 4c. It is interesting to note that the pores can easily be seen in the surface view of the SEM image. Surface SEM image of sample A, showing pores by arrows is shown in Figure 4d. The average size of pores is found to be 0.2 µm, 0.15 µm and 0.075 µm for sample A, B and C, respectively. It has been found that the average size and the surface density of the pores decrease upon annealing. We have found that the sizes of pores are further decreased upon increasing the annealing temperature to 300 °C for sample D. In addition, we have observed that the size of nanoparticles increases with increasing the annealing temperature and they are of the order of 6 nm, 8 nm, 12 nm and 15 nm for samples A, B, C and D, respectively.

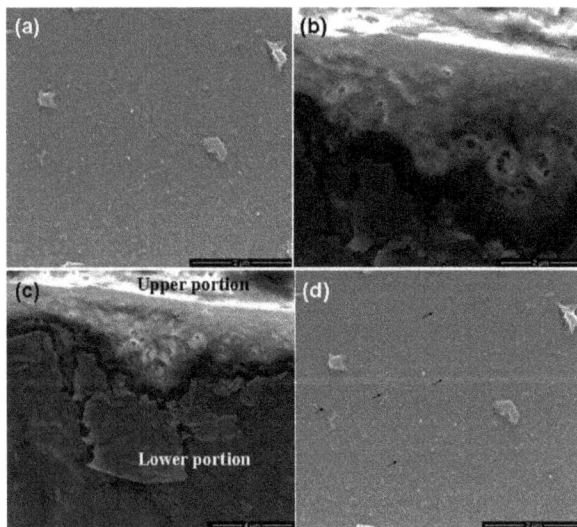

Figure 4. Scanning electron microscopy (SEM) images of as-deposited porous GaAs (sample A). (**a**) surface, (**b**) cross-section view, (**c**) cross-section view showing upper and lower portions, (**d**) pores shown by arrow on surface.

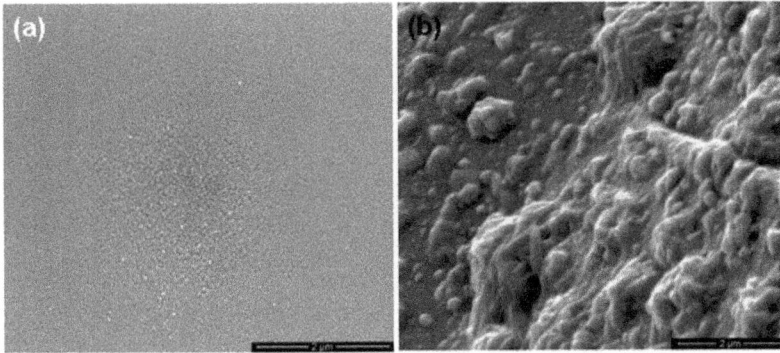

Figure 5. SEM images of 100 °C annealed porous GaAs (sample B). (**a**) surface, (**b**) cross-section view.

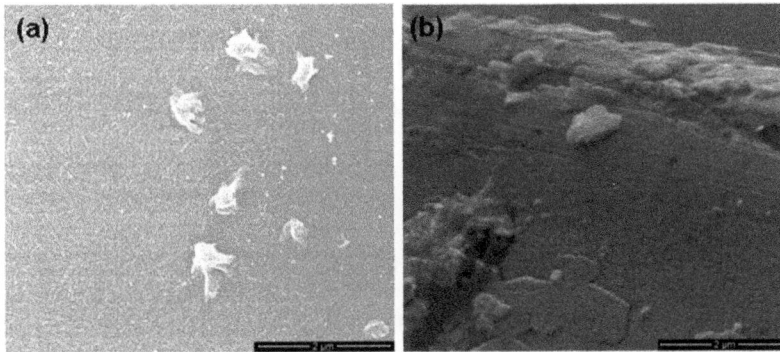

Figure 6. SEM images of 200 °C annealed porous GaAs (sample C). (**a**) surface, (**b**) cross-section view.

The observed decrease in the average size and surface density of the pores upon annealing can be understood as follows. It is well known that the pores are generally treated as defects in nanostructured films. It is observed in thin films that annealing removes the defects in the material. In the same way nanostructured films processed by DPF device were annealed and pores in these films were drastically reduced by annealing. This is possibly because of the process of rearrangement of GaAs to occupy the unfilled pores or due to agglomeration of nanoparticles outside of the pores caused by the annealing.

To ascertain the structure of the deposited material, X-ray diffraction (XRD) pattern of as-deposited and annealed samples has been investigated. XRD pattern of as-grown sample A and annealed samples are shown in Figure 7. The XRD pattern of as-grown sample indicates the presence of peaks at 2θ values of 27.3° and 45.3° corresponding to [111] and [220] diffraction planes of zinc blende GaAs (JCPDS File No. 14-450), respectively. The grain size has been calculated using Debye-Scherrer's Equation:

$$D = \frac{0.9\lambda}{\beta\cos\theta} \tag{1}$$

where D is the average grain size, λ is the wavelength, β is the full width at half maximum (FWHM) of the peak in radians, and θ is the angle of diffraction corresponding to the peak. The grain size is found to be ~10 nm and ~12 nm for [111] and [220] planes, respectively. Thus, the XRD results confirm the deposition of GaAs having nano-dimensional grains on the glass substrate. The XRD of the annealed samples has shown the peaks at the same position *viz.* 27.3° and 45.3° corresponding to [111] and [220] diffraction planes of GaAs, respectively. The grain sizes for the annealed samples are also of the same order as those obtained for the as-grown samples.

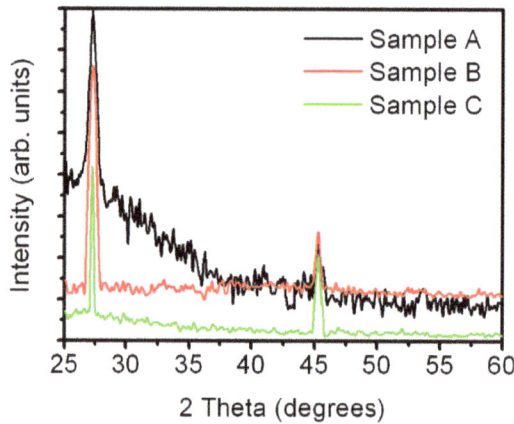

Figure 7. X-ray diffraction (XRD) pattern of as-deposited and rapid thermal annealing (RTA) porous GaAs.

The PL spectra of as-grown porous GaAs reveal the emission peaks at 511 nm and 817 nm. This emission suggests possible applications of the porous GaAs in sensors and light emitting devices. The changes in PL spectra with rapid thermal annealing were further studied. The PL spectra of the as-deposited porous GaAs along with the annealed samples are shown in Figure 8. For sample B which is annealed at 100 °C the above peaks are red-shifted to 518 nm and 824 nm. A similar red shift was observed when samples C and D were annealed further at 200 °C and 300 °C. In particular, the peaks were further red shifted to 525 nm and 831 nm for sample C and 538 nm and 844 nm for sample D.

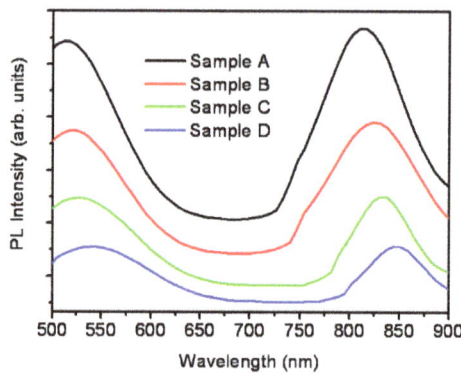

Figure 8. PL spectra of as-deposited and RTA porous GaAs.

The observed red shift in PL peaks upon annealing is likely due to a decrease in the surface density of the pores. The peaks observed in infra-red region at 817, 824, 831, and 844 nm are band-edge emission of GaAs having blue shift of ~50, 43, 36, and 23 nm, respectively, from the 867 nm peak of infrared emission originating from bulk GaAs [55].

The PL peaks observed in the visible region at 511, 518, 525, and 538 nm represent green emission from porous GaAs [49,53,56–58]. In literature, the green emission has been observed in PL spectra of porous GaAs due to the presence of nanocrystallites having size of the order of 7 nm [49].

We have also have estimated the size of nanocrystallites from the green emission peak using the effective mass theory. Assuming the infinite potential barriers, the energy gap E of GaAs nanocrystallites confined in three-dimensions should vary as [55],

$$E = E_g + \frac{h^2}{2d^2}\left(\frac{1}{m_e^*} + \frac{1}{m_h^*}\right) \tag{2}$$

where E_g is the energy bandgap of bulk GaAs, d is the diameter of nanocrystallites, while m_e^* and m_h^* are the effective masses of the electrons and holes, respectively. At 300 K, $E_g = 1.425$ eV, $m_e^* = 0.063 m_0$ and $m_h^* = 0.53 m_0$ [55], where m_0 is the electron mass in vacuum. The green PL peaks observed at 511, 518, 525, and 538 nm corresponds to the energies ~2.426, 2.394, 2.362, and 2.304 eV, respectively. From these data the value of d was estimated to be ~5 nm in all the four cases. Thus, the porous GaAs fabricated through the present technique possesses nanocrystallites having sizes of the order of 5 nm which give rise to green emission peak in PL spectra.

Transmission measurements give optical losses in porous GaAs. It is observed in the present study that there is a decrease in the percentage of transmission from 78% to 45% with increasing the annealing temperature in the shorter-wavelength range from 350 to 900 nm (Figure 9). On the other hand, in the longer-wavelength range from 900 to 1100 nm the percentage transmission is in the range of 80%–90% for all the four samples. This variation in the transmission spectra is due to the defect-related states present in the porous nano-GaAs. The increase in the transmission of GaAs in the infra-red region due to defect states has also been reported in the literature [59,60]. The high percentage transmission of the deposited and subsequently subjected to rapid thermal annealing (RTA) GaAs can find potential applications in transmission-type photocathodes [61]. The RTA process can, thus, be used for suppression the pore defects and for tailoring the pore, and consequently, the nanoparticle sizes.

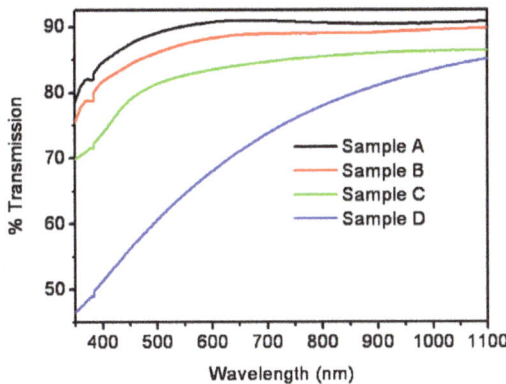

Figure 9. Transmission spectra of as-deposited and RTA porous GaAs.

The fabricated porous nano-GaAs has emission in visible range which suggests that the fabricated porous GaAs can be easily used in making nano-optoelectronic devices in the visible range. Moreover, the porous GaAs produced in the present experiment is the first time achievement by the physical based plasma method. Thus, the obtained porous nano-GaAs possess all qualities required for device fabrication. In addition, due to use of plasma based physical method the possibility of presence of contaminations is low which subsequently render the porous nano-GaAs more suitable for device fabrication. The porous GaAs fabricated earlier by Salehi and Kalantari [62] has already been used for making CO and NO gas sensors.

3. Experimental Section

A schematic of the DPF device with the modified anode and other modification is shown in Figure 10. The DPF device used is a 3.3 kJ Mather type device [63], powered by 30 μF, 15 kV energy storage capacitor. The anode of the DPF device has been modified such that a disc/pellet of a target material can be fixed at the top of it. The nanofabrication of different materials has been done on different substrates in the modified DPF device through high fluence of ions of the deposited material. The cleaned substrates are mounted on a perspex holder having a threaded hole at the center for screwing in the vertically moveable brass rod. The whole assembly is inserted from the top of the plasma chamber and its distance from the anode top is adjusted from outside the plasma chamber using a brass rod. The substrates are kept at room temperature. An aluminum (Al) shutter is also introduced from the top of the plasma chamber using another moveable brass rod. The shutter is placed in between the anode top and the substrates in order to protect the substrates from the impact of ions produced by unfocused plasmas. The chamber is evacuated using a rotary pump and then argon gas is introduced into the chamber as a working gas. The plasma chamber is flushed with argon gas several times to maintain the inert atmosphere. It has been established earlier [64] that an optimum argon gas pressure of 80 Pa is required for obtaining focused plasmas.

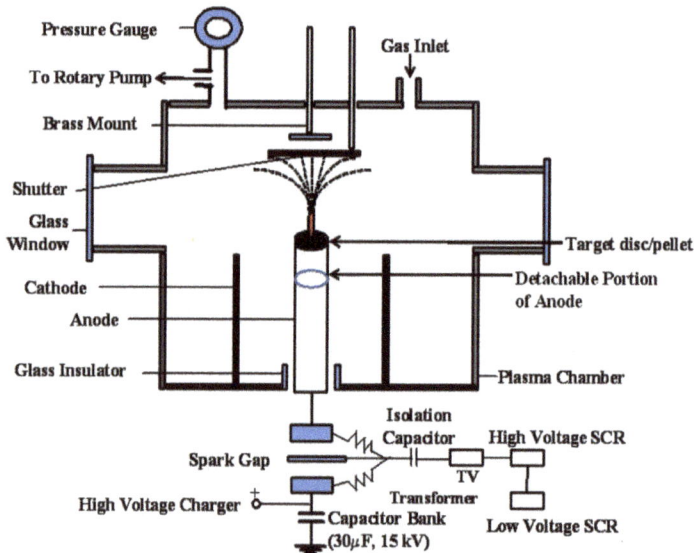

Figure 10. Schematic of modified dense plasma focus device.

In our experiments, we have also observed that a good focus is obtained at optimum argon gas pressure of 80 Pa in the chamber. The generation of the focused plasmas is observed as a sharp peak in the voltage signal which is recorded on a digital storage oscilloscope (Tektronix TDS 1002, Bangalore, India). The plasma has a density of the order of 10^{26} m^{-3}, temperature 1–2 keV and only lasts ~100 ns in the focus phase. At this stage, approximately 10^{17} ions (argon ions and ions of the deposited material) per shot are produced. The focused plasma ionizes the material disc/pellet surface and ions of the ablated material move vertically upwards in a fountain-like shape in the post-focus phase and then get deposited on the substrates. More discussions of the unique features of DPF devices and other methods of energetic ion deposition can be found elsewhere [65]. We have used one, two and three focused shots for nanofabrication of III–V semiconductors. The substrates were placed at 4.0 cm and 5.0 cm distance from the top of the anode.

4. Conclusions

The advantages of the modified DPF device are highlighted over the other deposition methods. In particular, this paper discussed the deposition of III–V semiconductor nanostructures using the high fluence ions generated in the modified DPF device. The brief results of the GaN and GaAs nanostructures fabricated using the modified DPF device have been discussed. The details of the first time fabrication of the porous nano-GaAs using modified DPF device are presented. It is found that RTA technique can be used to tailor the pore sizes in the porous nano-GaAs. The presence of nanograins in the porous GaAs samples has been confirmed from the XRD pattern. Furthermore, porous GaAs show strong PL emission in visible and infra-red regions along with high transmission. The obtained change in optical properties of porous GaAs on annealing is consistent with the changes observed in morphology. The porous GaAs fabricated and subsequently annealed in this work is promising for applications in visible optoelectronic devices and transmission type photocathodes.

Acknowledgments: Onkar Mangla is Guest faculty at Hindu College, Delhi University.

Author Contributions: Onkar Mangla and Savita Roy conducted the study and analyzed the data. All the three authors drafted the manuscript. Kostya (Ken) Ostrikov contributed to data and results interpretation and provided valuable suggestion for improvement of the manuscript. All authors approved the final version.

Conflicts of Interest: The authors declare no conflict of interest.

References

1. Surla, V.; Ruzic, D. High-energy density beams and plasmas for micro- and nano-texturing of surfaces by rapid melting and solidification. *J. Phys. D* **2011**, *44*. [CrossRef]
2. Rawat, R.S. High-Energy-Density Pinch Plasma: A Unique Nonconventional Tool for Plasma Nanotechnology. *IEEE Trans. Plasma Sci.* **2013**, *41*, 701–715. [CrossRef]
3. Umar, Z.A.; Rawat, R.S.; Tan, K.S.; Kumar, A.K.; Ahmad, R.; Hussain, T.; Kloc, C.; Chen, Z.; Shen, L.; Zhang, Z. Hard TiCx/SiC/a-C:H nanocomposite thin films using pulsed high energy density plasma focus device. *Nucl. Instrum. Methods Phys. Res. B* **2013**, *301*, 53–61. [CrossRef]
4. Khalid, M.; Mujahid, M.; Khan, A.N.; Rawat, R.S. Dip Coating of Nano Hydroxyapatite on Titanium Alloy with Plasma Assisted γ-Alumina Buffer Layer: A Novel Coating Approach. *J. Mater. Sci. Technol.* **2013**, *29*, 557–564. [CrossRef]
5. Umar, Z.A.; Rawat, R.S.; Ahmad, R.; Kumar, A.K.; Wang, Y.; Hussain, T.; Chen, Z.; Shen, L.; Zhang, Z. Mechanical properties of Al/a-C nanocomposite thin films synthesized using a plasma focus device. *Chin. Phys. B* **2014**, *23*, 1–6. [CrossRef]
6. Hussain, A.; Rawat, R.S.; Ahmad, R.; Hussain, T.; Umar, Z.A.; Ikhlaq, U.; Chen, Z.; Shen, L. A study of structural and mechanical properties of nano-crystalline tungsten nitride film synthesis by plasma focus. *Radiat. Eff. Def. Solids* **2015**, *170*, 73–83. [CrossRef]
7. Srivastava, M.P. Plasma Route to Nanosciences and Nanotechnology Frontiers. *J. Plasma Fusion Res.* **2009**, *8*, 512–516.
8. Singh, W.P.; Srivastava, M.P.; Roy, S. Nanoparticles and Nanostructured Cobalt Deposition Using Dense Plasma Focus Device. *J. Plasma Fusion Res.* **2009**, *8*, 526–529.
9. Singh, W.P.; Roy, S.; Srivastava, M.P. Formation of iron nanoparticles on quartz substrate using dense plasma focus device. *J. Phys. Conf. Ser.* **2010**, *208*. [CrossRef]
10. Devi, N.B.; Roy, S.; Srivastava, M.P. Deposition of aluminium nanoparticles using dense plasma focus device. *J. Phys. Conf. Ser.* **2010**, *208*. [CrossRef]
11. Naorem, B.D.; Roy, S.; Malhotra, Y.; Srivastava, M.P. Fabrication of gold nanostructures and studies of their morphological and surface plasmonic properties. *Plasmonics* **2013**, *8*, 1273–1278. [CrossRef]

12. Malhotra, Y.; Srivastava, M.P. AFM, XRD and Optical Studies of Silver Nanostructures Fabricated under Extreme Plasma Conditions. *J. Phys. Conf. Ser.* **2014**, *511*. [CrossRef]
13. Srivastava, M.P.; Naorem, B.D. Surface Plasmon Properties of Silver Nanostructures Fabricated using Extremely Non-Equilibrium Hot and Dense Plasma. *Adv. Mater. Res.* **2015**, *1110*, 226–230. [CrossRef]
14. Malhotra, Y.; Roy, S.; Srivastava, M.P.; Kant, C.R.; Ostrikov, K. Extremely non-equilibrium synthesis of luminescent zinc oxide nanoparticles through energetic ion condensation in a dense plasma focus device. *J. Phys. D* **2009**, *42*, 155202–155208. [CrossRef]
15. Malhotra, Y.; Roy, S.; Srivastava, M.P. Deposition and surface characterization of nanoparticles of zinc oxide using dense plasma focus device in nitrogen atmosphere. *J. Phys. Conf. Ser.* **2010**, *208*. [CrossRef]
16. Mangla, O.; Srivastava, M.P. GaN nanostructures by hot dense and extremely non-equilibrium plasma and their characterizations. *J. Mater. Sci.* **2013**, *48*, 304–310. [CrossRef]
17. Mangla, O.; Roy, S.; Srivastava, M.P. Synthesis and characterization of gallium arsenide nanostructured film for optoelectronic applications. *Adv. Sci. Eng. Med.* **2014**, *6*, 1200–1204. [CrossRef]
18. Mangla, O.; Roy, S. A study on aberrations in energy band gap of quantum confined gallium arsenide spherical nanoparticles. *Mater. Lett.* **2015**, *143*, 48–50. [CrossRef]
19. Srivastava, A.; Nahar, R.K.; Sarkar, C.K.; Singh, W.P.; Malhotra, Y. Study of hafnium oxide deposited using Dense Plasma Focus machine for film structure and electrical properties as a MOS device. *Microelectron. Reliab.* **2011**, *51*, 751–755. [CrossRef]
20. Mangla, O.; Srivastava, A.; Malhotra, Y.; Ostrikov, K. Lanthanum oxide nanostructured films synthesized using hot dense and extremely non-equilibrium plasma for nanoelectronic device applications. *J. Mater. Sci.* **2014**, *49*, 1594–1605. [CrossRef]
21. Mangla, O.; Srivastava, A.; Malhotra, Y.; Ostrikov, K. Metal-insulator-metal capacitors based on lanthanum oxide high-κ dielectric nanolayers fabricated using dense plasma focus device. *J. Vac. Sci. Technol. B* **2014**, *32*. [CrossRef]
22. Sinha, G.; Panda, S.K.; Mishra, P.; Ganguli, D.; Chaudhuri, S. Gallium nitride quantum dots in nitrogen-bonded silica gel matrix. *J. Phys. Condens. Matter* **2007**, *19*. [CrossRef]
23. Hirasawa, M.; Ichikawa, N.; Egashira, Y.; Honma, I.; Komiyama, H. Synthesis of GaAs nanoparticles by digital radio frequency sputtering. *Appl. Phys. Lett.* **1995**, *67*, 3483–3485. [CrossRef]
24. Shetty, S.; Kesaria, M.; Ghatak, J.; Shivaprasad, S.M. The Origin of Shape, Orientation, and Structure of Spontaneously Formed Wurtzite GaN Nanorods on Cubic Si(001) Surface. *Cryst. Growth Des.* **2013**, *13*, 2407–2412. [CrossRef]
25. Davydok, A.; Biermanns, A.; Pietsch, U.; Grenzer, J.; Paetzelt, H.; Gottschalch, V.; Bauer, J. Submicron resolution X-ray diffraction from periodically patterned GaAs nanorods grown onto Ge[111]. *Phys. Stat. Solidi A* **2009**, *206*, 1704–1708. [CrossRef]
26. Li, H.D.; Zhang, S.L.; Yang, H.B.; Zou, G.T.; Yang, Y.Y.; Yue, K.T.; Wu, X.H.; Yan, Y. Raman spectroscopy of nanocrystalline GaN synthesized by arc plasma. *J. Appl. Phys.* **2002**, *91*, 4562–4567. [CrossRef]
27. Okamoto, S.; Kanemitsu, Y.; Min, K.S.; Atwater, H.A. Photoluminescence from GaAs nanocrystals fabricated by Ga$^+$ and As$^+$ co-implantation into SiO$_2$ matrices. *Appl. Phys. Lett.* **1998**, *73*, 1829–1831. [CrossRef]
28. Wang, Y.D.; Chua, S.J.; Tripathy, S.; Sander, M.S.; Chen, P.; Fonstad, C.G. High optical quality GaN nanopillar arrays. *Appl. Phys. Lett.* **2005**, *86*. [CrossRef]
29. DeJarld, M.; Shin, J.C.; Chern, W.; Chanda, D.; Balasundaram, K.; Rogers, J.A.; Li, X. Formation of High Aspect Ratio GaAs Nanostructures with Metal-Assisted Chemical Etching. *Nano Lett.* **2011**, *11*, 5259–5263. [CrossRef] [PubMed]
30. Wierzbicka, A.; Zytkiewicz, Z.R.; Kret, S.; Borysiuk, J.; Dluzewski, P.; Sobanska, M.; Klosek, K.; Reszka, A.; Tchutchulashvili, G.; Cabaj, A.; et al. Influence of substrate nitridation temperature on epitaxial alignment of GaN nanowires to Si(111) substrate. *Nanotechnology* **2013**, *24*. [CrossRef] [PubMed]
31. Jahn, U.; Lahnemann, J.; Pfuller, C.; Brandt, O.; Breuer, S.; Jenichen, B.; Ramsteiner, M.; Geelhaar, L.; Riechert, H. Luminescence of GaAs nanowires consisting of wurtzite and zinc-blende segments. *Phys. Rev. B* **2012**, *85*. [CrossRef]
32. Ghosh, C.; Pal, S.; Goswami, B.; Sarkar, P. Theoretical Study of the Electronic Structure of GaAs Nanotubes. *J. Phys. Chem. C* **2007**, *111*, 12284–12288. [CrossRef]
33. Hemmingsson, C.; Pozina, G.; Khromov, S.; Monemar, B. Growth of GaN nanotubes by halide vapor phase epitaxy. *Nanotechnology* **2011**, *22*. [CrossRef] [PubMed]

34. Hwang, G.; Dockendorf, C.; Bell, D.; Dong, L.; Hashimoto, H.; Poulikakos, D.; Nelson, B. 3-D InGaAs/GaAs Helical Nanobelts for Optoelectronic Devices. *Inter. J. Optomech.* **2008**, *2*, 88–103. [CrossRef]
35. Yu, R.; Dong, L.; Pan, C.; Niu, S.; Liu, H.; Liu, W.; Chua, S.; Chi, D.; Wang, Z.L. Piezotronic Effect on the Transport Properties of GaN Nanobelts for Active Flexible Electronics. *Adv. Mater.* **2012**, *24*, 3532–3537. [CrossRef] [PubMed]
36. Moon, J.Y.; Kwon, H.Y.; Shin, M.J.; Choi, Y.J.; Ahn, H.S.; Chang, J.H.; Yi, S.N.; Yun, Y.J.; Ha, D.H.; Park, S.H. Growth behavior of GaN nanoneedles with changing HCl/NH$_3$ flow ratio. *Mater. Lett.* **2009**, *63*, 2695–2697. [CrossRef]
37. Chuang, L.C.; Sedgwick, F.G.; Chen, R.; Ko, W.S.; Moewe, M.; Ng, K.W.; Tran, T.T.D.; Chang-Hasnain, C. GaAs-Based Nanoneedle Light Emitting Diode and Avalanche Photodiode Monolithically Integrated on a Silicon Substrate. *Nano Lett.* **2011**, *11*, 385–390. [CrossRef] [PubMed]
38. Kim, H.M.; Kim, D.S.; Park, Y.S.; Kim, D.Y.; Kang, T.W.; Chung, K.S. Growth of GaN Nanorods by a Hydride Vapor Phase Epitaxy Method. *Adv. Mater.* **2002**, *14*, 991–993. [CrossRef]
39. Yang, Y.G.; Ma, H.L.; Xue, C.S.; Zhuang, H.Z.; Ma, J.; Hao, X.T. Preparation and properties of GaN nanostructures by post-nitridation technique. *Phys. B* **2003**, *334*, 287–291. [CrossRef]
40. Hung, J.; Lee, S.C.; Chia, C.T. The structural and optical properties of gallium arsenic nanoparticles. *J. Nanopart. Res.* **2004**, *6*, 415–419. [CrossRef]
41. Nayak, J.; Mythili, R.; Vijayalakshmi, M.; Sahu, S.N. Size quantization effect in GaAs nanocrystals. *Phys. E* **2004**, *24*, 227–233. [CrossRef]
42. Czaban, J.A.; Thompson, D.A.; LaPierre, R.R. GaAs Core-Shell Nanowires for Photovoltaic Applications. *Nano Lett.* **2009**, *9*, 148–154. [CrossRef] [PubMed]
43. Ng, D.K.T.; Tan, L.S.; Hong, M.H. Synthesis of GaN nanowires on gold-coated substrates by pulsed laser ablation. *Curr. Appl. Phys.* **2006**, *6*, 403–406. [CrossRef]
44. Lei, M.; Yang, H.; Li, P.G.; Tang, W.H. Synthesis of GaN nanowires on gold-coated SiC substrates by novel pulsed electron deposition technique. *Appl. Surf. Sci.* **2008**, *254*, 1947–1952. [CrossRef]
45. Dinh, L.N.; Hayes, S.; Saw, C.K.; McLean, W., II; Balooch, M.; Reimer, J.A. GaAs nanostructures and films deposited by a Cu-vapor laser. *Appl. Phys. Lett.* **1999**, *75*, 2208–2210. [CrossRef]
46. Duan, X.; Wang, J.; Lieber, C.M. Synthesis and optical properties of gallium arsenide nanowires. *Appl. Phys. Lett.* **2000**, *76*, 1116–1118. [CrossRef]
47. Yusa, G.; Sakaki, H. Trapping of photogenerated carriers by InAs quantum dots and persistent photoconductivity in novel GaAs/n-AlGaAs field-effect transistor structures. *Appl. Phys. Lett.* **1997**, *70*, 345–347. [CrossRef]
48. Page, H.; Becker, C.; Robertson, A.; Glastre, G.; Ortiz, V.; Sirtori, C. 300 K operation of a GaAs-based quantum-cascade laser at $\lambda \approx 9$ μm. *Appl. Phys. Lett.* **2001**, *78*, 3529–3531. [CrossRef]
49. Belogorokhov, A.I.; Gavrilov, S.A.; Belogorokhov, I.A. Structural and optical properties of porous gallium arsenide. *Phys. Status Solidi C* **2005**, *2*, 3491–3494. [CrossRef]
50. Grym, J.; Nohavica, D.; Vanis, J.; Piksova, K. Preparation of nanoporous GaAs substrates for epitaxial growth. *Phys. Status Solidi C* **2012**, *9*, 1531–1533. [CrossRef]
51. Oskam, G.; Natarajan, A.; Searson, P.C.; Ross, F.M. The formation of porous GaAs in HF solution. *Appl. Surf. Sci.* **1997**, *119*, 160–168. [CrossRef]
52. Beji, L.; Sfaxi, L.; Ismail, B.; Missaoui, A.; Hassen, F.; Maaref, H.; Ouada, H.B. Visible photoluminescence in porous GaAs capped by GaAs. *Phys. E* **2005**, *25*, 636–642. [CrossRef]
53. Dmitruk, N.; Kutovyi, S.; Dmitruk, I.; Simkiene, I.; Sabataityte, J.; Berezovska, N. Morphology, Raman scattering and photoluminescence of porous GaAs layers. *Sens. Actuators B* **2007**, *126*, 294–300. [CrossRef]
54. Srinivasan, R.; Ramachandran, K. Synthesis and thermal diffusion of nanostructured porous GaAs. *Cryst. Res. Technol.* **2008**, *43*, 953–958. [CrossRef]
55. Madelung, O. *Semiconductors: Group IV Elements and III–V Compounds*; Springer: Berlin, Germany, 1991; pp. 101–103.
56. Schmuki, P.; Lockwood, D.J.; Labbe, H.J.; Fraser, J.W. Visible photoluminescence from porous GaAs. *Appl. Phys. Lett.* **1996**, *69*, 1620–1622. [CrossRef]
57. Schmuki, P.; Erickson, L.E.; Lockwood, D.J.; Fraser, J.W.; Champion, G.; Labbe, H.J. Formation of visible light emitting porous GaAs micropatterns. *Appl. Phys. Lett.* **1998**, *72*, 1039–1041. [CrossRef]

58. Lockwood, D.J.; Schmuki, P.; Labbe, H.J.; Fraser, J.W. Optical properties of porous GaAs. *Phys. E* **1999**, *4*, 102–110. [CrossRef]

59. Skolnick, M.S.; Reed, L.J.; Pitt, A.D. Photoinduced quenching of infrared absorption nonuniformities of large diameter GaAs crystals. *Appl. Phys. Lett.* **1984**, *44*, 447–449. [CrossRef]

60. Rudolph, P. Non-stoichiometry related defects at the melt growth of semiconductor compound crystals—A review. *Cryst. Res. Technol.* **2003**, *38*, 542–554. [CrossRef]

61. Yamamoto, N.; Nakanishi, T.; Mano, A.; Nakagawa, Y.; Okumi, S.; Yamamoto, M.; Konomi, T.; Jin, X.; Ujihara, T.; Takeda, Y.; *et al.* High brightness and high polarization electron source using transmission photocathode with GaAs-GaAsP superlattice layers. *J. Appl. Phys.* **2008**, *103*. [CrossRef]

62. Salehi, A.; Kalantari, D.J. Characteristics of highly sensitive Au/porous-GaAs Schottky junctions as selective CO and NO gas sensors. *Sens. Actuators B Chem.* **2007**, *122*, 69–74. [CrossRef]

63. Mather, J.W. Investigation of the high energy acceleration mode in the coaxial gun. *Phys. Fluids* **1964**, *7*, S28–S34. [CrossRef]

64. Rawat, R.S.; Srivastava, M.P.; Tandon, S.; Mansingh, A. Crystallization of an amorphous lead zirconate titanate thin film with a dense-plasma-focus device. *Phys. Rev. B* **1993**, *47*, 4858–4862. [CrossRef]

65. Ostrikov, K.; Neyts, E.C.; Meyyappan, M. Plasma nanoscience: From nano-solids in plasmas to nano-plasmas in solids. *Adv. Phys.* **2013**, *62*, 113–224. [CrossRef]

treatment homogeneity to normally modify the defects. The resistive switching of RRAMs might be adjusted as well. However, the effects of plasma treatment on the switching properties of RRAMs are scarcely investigated [6,20]. We therefore fabricated single ZnO NW–based RRAMs with a Ti/ZnO NW/Ti structure. Argon plasma treatment was also applied to the ZnO NW to evaluate the effects of plasma treatment on the switching properties of the RRAMs.

2. Results and Discussion

The switching properties are critically important for the practical applications of the RRAMs. We hereby evaluate the effects of plasma treatment on the switching properties of the ZnO NW–based memory according to the homogeneity and reproducibility of the switching parameters, the data retention and the embedded self-rectification capability.

2.1. Enhanced Homogeneity and Reproducibility

Storage properties of a RRAM are criterions to estimate the effectiveness of plasma treatment. Figure 1 shows the voltage-biased current-voltage (I–V) curves of the plasma-treated and untreated memory cells at every 10 switching cycles (marked with blue, black and red symbols). To improve the readability of the I–V curves, the curves from both the plasma-treated samples and the untreated ones are shown on log-log scale (Figure 1a,b) and are respectively accompanied with curves on normally log-linear scale (Figure 1c,d). Both memory cells exhibit bipolar resistive switching with decreased set and reset voltages by plasma treatment. In addition, the repeatability of switching behavior is significantly enhanced by plasma treatment. To get insights into the switching mechanism, the I–V curves on log-log scale were studied. Each I–V curve during the set period is composed of three portions with different slopes (the Ohmic region with $I \sim V$, the Child's square-law region with $I \sim V^2$, and the exponentially increased current region with $I \sim V^a$), showing a typical feature of space-charge-limited conduction (SCLC) [21,22]. The oxygen vacancy-assisted filaments and metal-dominated filaments could be closely associated with the electrical switching between the HRS and the LRS [12,23,24]. Figure 2 shows the temperature-dependent resistance of the LRS memory cells. The decreasing resistance upon the increased temperature is in good agreement with a semiconductor conduction but denies a metallic conduction [6]. We thus deduce that the rupture and formation of oxygen vacancy-assisted filaments should be respectively responsible for the reset and the set process. The formation of TiO_x at the Ti/ZnO interface is inevitable in this work due to a much higher enthalpy of the formation for TiO_2 (-944 kJ/mol) than that for ZnO (-350 kJ/mol) [10]. During the revisable resistance switching, the oxygen atoms in the TiO_x would also migrate back and forth between the TiO_x and the ZnO upon the applied electric fields to facilitate the rupture and formation of the conductive filaments.

The reversible switching tests were subsequently carried out for another 100 cycles to estimate the endurance of the memory cells. Figure 3a shows the endurance of the memory cell without plasma treatment, and the plasma-treated one is shown in Figure 3b. At the first 30 switching cycles, the HRS of the untreated memory is unstable with fluctuations in resistance. It gets reproducible during the left 70 cycles. The LRS of the untreated memory starts with acceptably reproducible resistances but its repeatability becomes worse after 80 switching cycles. However, both the HRS and the LRS of the plasma-treated memory are reproducible with ignorable fluctuations in resistance over the 100 switching cycles, though the resistance ratio between the HRS and the LRS is reduced by plasma treatment.

For the practical applications of the RRAMs, the distributions of the set and reset voltages (also known as switching voltages) from different cells should be as narrow as possible to ensure operation reliability. The distributions of the switching voltages were then analyzed and shown in Figure 4. For the untreated ZnO NW, the reset voltages spanning from -3 V to -31 V and the set voltages spanning from 3 to 22 V can be observed without obvious peaking voltages. However, for the plasma-treated ZnO NW memory, both the reset voltages and the set voltages present normal distributions and respectively span from -1.5 to -10 V and from 0.5 to 8 V with peaking at -7 and 3 V, which indicates that the switching voltages decrease by plasma treatment with an increase in homogeneity.

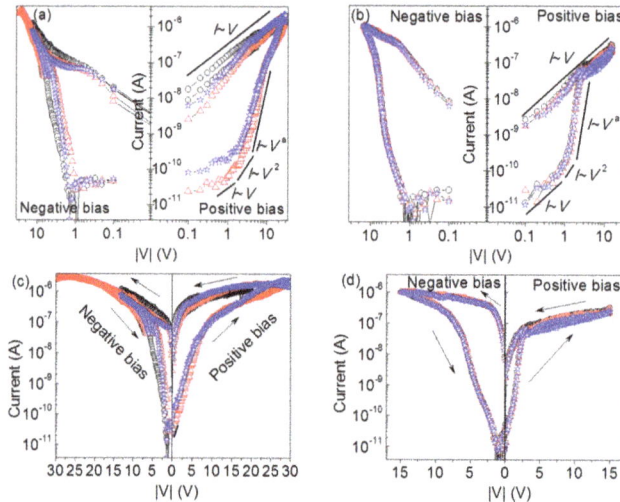

Figure 1. Reproducible voltage-biased current-voltage (*I–V*) curves of the ZnO nanowires (ZnO NW)–based memories (**a,c**) without and (**b,d**) with argon plasma treatment. Blue (☆), black (O) and red (△) symbols respectively represent the 8th, 18th and 28th switching cycles.

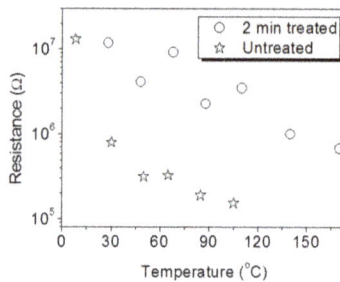

Figure 2. Temperature-dependent resistance of the low resistance state (LRS) memories with and without plasma treatment.

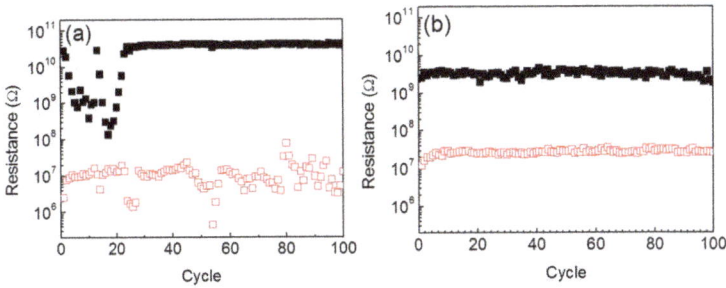

Figure 3. Endurance tests of the ZnO NW–based memories (**a**) without and (**b**) with argon plasma treatment.

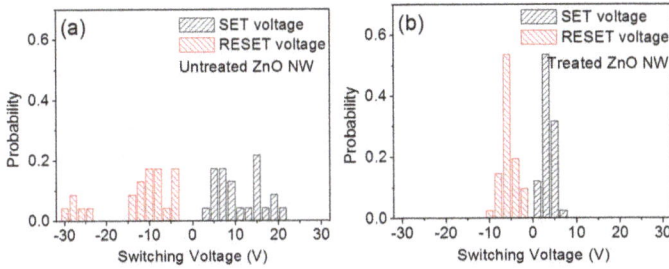

Figure 4. Distributions of switching voltages of the ZnO NW–based memories (**a**) without and (**b**) with argon plasma treatment.

2.2. Improved Data Retention

To further clarify the storage performance, the data retention of the memories was evaluated as shown in Figure 5. Plasma treatment reduces the resistance ratio between the two binary storage states but extends the data retention from less than one year to over 10 years by stabilizing the LRS resistance.

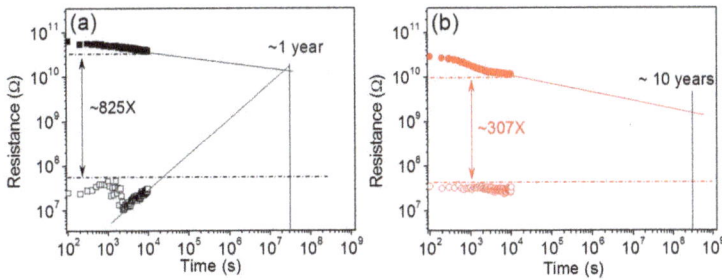

Figure 5. Data retention of the ZnO NW–based memories (**a**) without and (**b**) with argon plasma treatment.

2.3. Self-Rectification of the RRAM Cells

With the demand of ultra-high density integration from the memory industry, crossbar arrays as a feasible architecture for future RRAMs have attracted much attention in recent years [25–27]. However, the sneak-path issue is a key challenge for this scheme, since the read-out sense margin could be reduced by the undesired leakage current through the unselected memory cells. To solve this problem, the *I–V* curve of the LRS RRAM cell should be non-linear, which could be realized by the integration of a memory resistor and a rectification diode [28]. Except for the almost symmetric *I–V* characteristic shown in Figure 1, the asymmetric *I–V* characteristic could also be observed for the ZnO nanowires grown with ~4 nm gold thin films as shown in Figure 6a. The current ratio under the positive bias and the negative bias may reach 10,000, indicating an integrated self-rectification in the memory cell. When it comes to the switching mechanism, the *I–V* curves were replotted on log-log scale as shown in Figure 6b. An Ohmic region and a Child's law region form the *I–V* curves of the LRS RRAM cell, which is in agreement with the SCLC mechanism [21,22]. However, the SCLC mechanism does not dominate the conduction in the HRS, as a better linear fitting of $\sqrt{V} \sim \ln(I)$ indicates a Schottky emission dominant conduction [29,30]. Therefore, there should be a Schottky barrier at the Ti/ZnO interface. As we know, the rectification is associated with the interface status. Figure 7 shows the scanning electron microscopy (SEM) images of the as-grown ZnO nanowires with different thicknesses of gold catalyst. Compared with the thin (~2 nm) gold catalyzed nanowires in a lower density with a smaller diameter (see Figure 7a), the thick (~4 nm) gold catalyst produces nanowires in a higher density with a greater diameter (see Figure 7b). Additionally, there are some leaf-like species at the root of the thick (~4 nm) gold catalyzed ZnO nanowires to partially protect the nanowire's root from the plasma treatment (see Figure 7b). Considering the higher density, the protection effects for the root of the thick gold catalyzed nanowires would be further enhanced. The inhomogeneous interfacial treatment on the two terminals of the thick gold catalyzed nanowires would thus be obtained, instead of a much more uniform treatment throughout the whole nanowire with thin gold catalyst. The gold catalyst normally guides the growth of nanowires during the vapor-liquid-solid synthesis process and could hardly have a direct association with the self-rectification. Consequently, the gold catalyst determines the morphology of the as-grown ZnO nanowires. The high density and the leaf-like species contribute to the inhomogeneous interfacial treatment effects and result in self-rectification as shown in Figure 6a.

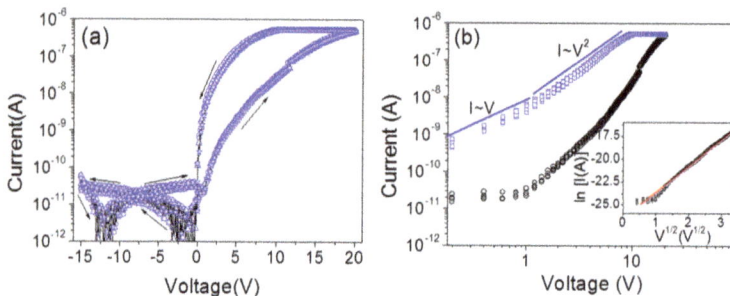

Figure 6. (a) Reproducible asymmetric *I–V* curves of single ZnO NW and (b) the *I–V* curves at positive bias on log-log scale with the inset fitting of $\sqrt{V} \sim \ln(I)$ for the high resistance state (HRS).

Figure 7. Cross-sectional scanning electron microscope (SEM) images of the ZnO NWs on the silicon substrates with gold thicknesses of (**a**) ~2 nm and (**b**) ~4 nm.

2.4. Defects in the ZnO NWs

Investigation on the defects is helpful for a better understanding of the reversible switching of a RRAM. Figure 8 shows the room-temperature photoluminescence (PL) spectra of the ZnO NWs on log-log scale in order to elucidate weak emissions. Each spectrum is composed of a ~380 nm centered UV emission and a broad visible emission spanning from ~430 nm to ~570 nm. The intensity ratios of the UV emission to the visible emission are about 21.06 and 6.75 for the untreated and the plasma-treated ZnO NWs, respectively, which means the argon plasma treatment introduces defects with visible emissions. We therefore de-convolute them into several symmetrical peaks as shown in Figure 8. For the untreated ZnO NWs, there is a weak ~470 nm centered emission in the visible region, indicating a good crystal structure with few defects. However, except for the ~470 nm centered emission, a ~510 nm centered emission could also be observed for the plasma-treated ZnO NWs, which suggests the increase of deep oxygen vacancy (V_o) by plasma treatment due to the agreement with the electron transition from the deep V_o level to the top of the valance band [31]. To verify the results, we underwent the experiment three times from three batches of samples. Similar results were observed. During the argon plasma treatment period, the oxygen atoms would be driven out of the ZnO surface by the collision of inert argon ions, which results in an oxygen-deficient surface with oxygen vacancies. The formation of oxygen vacancy-assisted conductive filaments triggers the set process to complete the resistance transition from the HRS to the LRS. If more oxygen vacancies are involved during this period, the uncertainty of the formation of the conductive filaments should be minimized as the distance between the adjacent V_os is reduced and the electron-hopping between them would be easier. The homogeneity and reproducibility of the resistive switching could thus be improved. Additionally, the increased V_o density may also effectively average out the stochastic movement of individual V_os to improve resistance stability as well as to prolong data retention [32]. Liu Ming also doped HfO_2 with nitrogen and found that the increased oxygen vacancies play quite similar roles [33]. However, the increased V_os generally act as dopants in oxide semiconductors to decrease the resistance and result in a slightly suppressed resistance ratio between the HRS and the LRS as what we observe in Figure 3b. As for the self-rectification in the plasma-treated RRAM cell, the effects of plasma treatment on the two back-to-back connected Schottky barriers at the Ti/ZnO contacts should be taken into account. The high V_o density usually results in the ZnO Fermi level pinning close to the defect level to increase the probability of electron tunneling by narrowing the region of positive space charge [34], and the rectification capability of the Ti/ZnO interface may be disabled. As a result, the plasma treatment weakens the rectification at one Ti/ZnO contact but indeed manifests the rectification capability at another Ti/ZnO contact. The self-rectification phenomenon could thus be observed as shown in Figure 6.

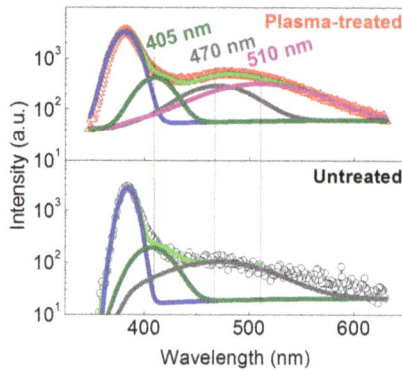

Figure 8. Room-temperature photoluminescence (PL) spectra of the ZnO NWs and their Gaussian components.

3. Experimental Section

Thin gold films (~2 nm and ~ 4 nm) were firstly evaporated onto the Si(100) substrate prior to an 800 °C annealing for 10 minutes to form discrete gold nano-particles as catalysts. The ZnO NWs were then grown by a vapor-liquid-solid (VLS) method at 975 °C with a mixture of ZnO and carbon powder as precursor. Subsequently, the argon plasma treatment was applied to the as-grown nanowires with 100 W at 120 Pa for 240 s. The morphologies of the as-grown nanowires are shown in Figure 7. Compared with the ZnO nanowires grown with ~4 nm gold films, the ZnO nanowires grown with ~2 nm gold films have smaller diameter in a lower density that ensures the plasma treatment be employed even to the root of the nanowires. While the root of the ZnO nanowires with ~4 nm gold films should be protected from plasma treatment by the coverage of surrounding materials. The inhomogeneous plasma treatment effects should be obtained as what we observed in Figure 6a. To fabricate a single ZnO NW–based RRAM cell, horizontal ZnO NW memory was selected because its fabrication procedure would be less complex compared with that for a vertical ZnO NW memory. Additionally, the considerations of switching mechanism would be the same for the two structures. Therefore, the nanowires were released from the substrates by ultrasonic vibration prior to the dispersion onto the SiO_2/Si(100) substrate. Two titanium electrodes spacing ~3 μm were subsequently sputtered and contacted with ZnO NW on its two terminals to form a memory cell as shown in Figure 9. To modulate the switching voltages and the resistance of the binary storage states, changing the plasma-treatment parameters or shortening the distance between the two electrodes might be tried in the consequent experiments.

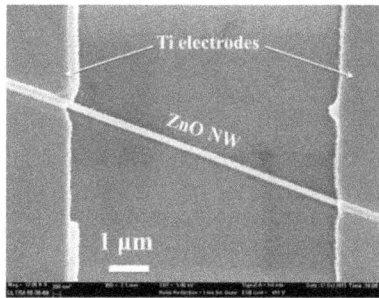

Figure 9. Top-view SEM images of the single ZnO NW–based resistive random access memory (RRAM) cell.

The *I–V* characteristics and the other switching properties of the RRAM cells were characterized using a semiconductor characterization system (4200-SCS; Keithley, OH, USA). To evaluate the effects of plasma treatment, we measured the untreated memories and the treated memories in the same environment to ensure the protons and the absorbed species have the same effects on the memory cells. The morphology of the ZnO NWs and the single ZnO NW–based memory cell was observed by a scanning electron microscopy (Ultra-55; Zeiss, Oberkochen, Germany). PL spectra with 325 nm as excitation wavelength were also investigated for a better understanding of the defects.

4. Conclusions

The effects of plasma treatment on the resistive switching of a single ZnO NW have been investigated. The plasma treatment enhances not only the homogeneity but also the reproducibility of the resistive switching by the plasma-produced oxygen vacancies to average out the stochastic movement of individual oxygen vacancies. The plasma treatment could also reduce switching voltages due to the increased oxygen vacancies facilitating resistive switching. Additionally, inhomogeneous plasma treatment on the two terminals may result in self-rectification in the memory cell by weakening the rectification of the Schottky barrier at one terminal.

Acknowledgments: This work was supported by the National Natural Science Foundation of China (Grant Nos. 61006003, 61306120, and 61440009), the Natural Science Foundation of Fujian Province (Grant Nos. 2015J01249 and 2010J05134), the Scientific Research Foundation for the Returned Overseas Chinese Scholars, State Education Ministry (Grant No. LXKQ201104), and the Science Foundation of Fujian Education Department (Grant No. JA15076).

Author Contributions: Yunfeng Lai designed the work and monitored the progress and he also took the responsibilities of data analysis, paper writing and editing. Wenbiao Qiu was responsible for the plasma treatment, device fabrication and characterization. Zecun Zeng was responsible for the synthesis of nanowires. Shuying Cheng, Jinling Yu and Qiao Zheng contributed to the useful discussion and suggestion.

Conflicts of Interest: The authors declare no conflict of interest.

1.	Pan, F.; Gao, S.; Chen, C.; Song, C.; Zeng, F. Recent progress in resistive random access memories: materials, switching mechanisms, and performance. *Mater. Sci. Eng. R* **2014**, *83*, 1–59. [CrossRef]

2.	Waser, R.; Dittmann, R.; Staikov, G.; Szot, K. Redox-based resistive switching memories-nanoionic mechanisms, prospects, and challenges. *Adv. Mater.* **2009**, *21*, 2632–2663. [CrossRef]

3.	Waser, R.; Aono, M. Nanoionics-based resistive switching memories. *Nat. Mater.* **2007**, *6*, 833–840. [CrossRef] [PubMed]

4.	Liang, L.; Li, K.; Xiao, C.; Fan, S.; Liu, J.; Zhang, W.; Xu, W.; Tong, W.; Liao, J.; Zhou, Y.; *et al.* Vacancy Associates-Rich Ultrathin Nanosheets for High Performance and Flexible Nonvolatile Memory Device. *J. Am. Chem. Soc.* **2015**, *137*, 3102–3108. [CrossRef] [PubMed]

5.   Qian, M.; Pan, Y.M.; Liu, F.Y.; Wang, M.; Shen, H.L.; He, D.W.; Wang, B.G.; Shi, Y.; Miao, F.; Wang, X.R. Tunable, ultralow-power switching in memristive devices enabled by a heterogeneous graphene-oxide interface. *Adv. Mater.* **2014**, *26*, 3275–3281. [PubMed]

6.   Lai, Y.F.; Xin, P.C.; Cheng, S.Y.; Yu, J.L.; Zheng, Q. Plasma enhanced multistate storage capability of single ZnO nanowire based memory. *Appl. Phys. Lett.* **2015**, *106*. [CrossRef]

7.   Yan, X.B.; Hao, H.; Chen, Y.F.; Li, Y.C.; Banerjee, W. Highly transparent bipolar resistive switching memory with In-Ga-Zn-O semiconducting electrode in In-Ga-Zn-O/Ga2O3/In-Ga-Zn-O structure. *Appl. Phys. Lett.* **2014**, *105*. [CrossRef]

8.   Kim, A.; Song, K.; Kim, Y.; Moon, J. All solution processed, fully transparent resistive memory devices. *ACS Appl. Mater. Interfaces* **2011**, *3*, 4525–4530. [CrossRef] [PubMed]

9.   Cao, X.; Li, X.M.; Gao, X.D.; Liu, X.J.; Yang, C.; Yang, R.; Jin, P. All-ZnO-based transparent resistance random access memory device fully fabricated at room temperature. *J. Phys. D* **2011**, *44*. [CrossRef]

10.   Chiang, Y.D.; Chang, W.Y.; Ho, C.Y.; Chen, C.Y.; Ho, C.H.; Lin, S.J.; Wu, T.B.; He, J.H. Single ZnO nanowire memory. *IEEE Trans. Electron Devices* **2011**, *58*, 1735–1740. [CrossRef]

11.   Lai, Y.; Wang, Y.; Cheng, S.; Yu, J. Defects and resistive switching of zinc oxide nanorods with copper addition grown by hydrothermal method. *J. Electron. Mater.* **2014**, *43*, 2676–2682. [CrossRef]

12.   Qi, J.; Olmedo, M.; Ren, J.J.; Zhan, N.; Zhao, J.Z.; Zheng, J.G.; Liu, J.L. Resistive switching in single epitaxial ZnO nanoislands. *ACS Nano* **2012**, *6*, 1051–1058. [CrossRef] [PubMed]

13.   Wang, Y.; Liu, Q.; Lu, H.B.; Long, S.B.; Wang, W.; Li, Y.T.; Zhang, S.; Lian, W.T.; Yang, J.H.; Liu, M. Improving the electrical performance of resistive switching memory using doping technology. *Chin. Sci. Bull.* **2012**, *57*, 1235–1240. [CrossRef]

14.   Guan, W.H.; Long, S.B.; Jia, R.; Liu, M. Nonvolatile resistive switching memory utilizing gold nanocrystals embedded in zirconium oxide. *Appl. Phys. Lett.* **2007**, *91*. [CrossRef]

15.   Wang, Y.; Liu, Q.; Long, S.; Wang, W.; Wang, Q.; Zhang, M.; Zhang, S.; Li, Y.; Zuo, Q.; Yang, J.; *et al.* Investigation of resistive switching in Cu-doped HfO2 thin film for multilevel non-volatile memory applications. *Nanotechnology* **2010**, *21*. [CrossRef]

16.   Chen, Y.T.; Chang, T.C.; Peng, H.K.; Tseng, H.C.; Huang, J.J.; Yang, J.B.; Chu, A.K.; Young, T.F.; Sze, S.M. Insertion of a Si layer to reduce operation current for resistive random access memory applications. *Appl. Phys. Lett.* **2013**, *102*. [CrossRef]

17.   Chen, Y.-S.; Lee, H.-Y.; Chen, P.-S.; Chen, W.-S.; Tsai, K.-H.; Gu, P.-Y.; Wu, T.-Y.; Tsai, C.-H.; Rahaman, S.Z.; Lin, Y.-D.; *et al.* Novel defects-trapping TaO$_X$/HfO$_X$ RRAM with reliable self-compliance, high nonlinearity, and ultra-low current. *IEEE Electron Device Lett.* **2014**, *35*, 202–204. [CrossRef]

18.   Kim, S.; Choi, S.; Lu, W. Comprehensive physical model of dynamic resistive switching in an oxide memristor. *ACS Nano* **2014**, *8*, 2369–2376. [CrossRef] [PubMed]

19.   Chen, Y.Y.; Goux, L.; Clima, S.; Govoreanu, B.; Degraeve, R.; Kar, G.S.; Fantini, A.; Groeseneken, G.; Wouters, D.J.; Jurczak, M. Endurance/retention trade-off on HfO2/metal cap 1T1R bipolar RRAM. *IEEE Trans. Electron Devices* **2013**, *60*, 1114–1121. [CrossRef]

20.   Chen, X.R.; Feng, J.; Bae, D. Drastic reduction of RRAM reset current via plasma oxidization of TaOx film. *Appl. Surface Sci.* **2015**, *324*, 275–279. [CrossRef]

21.   Kim, K.M.; Choi, B.J.; Shin, Y.C.; Choi, S.; Hwang, C.S. Anode-interface localized filamentary mechanism in resistive switching of TiO2 thin films. *Appl. Phys. Lett.* **2007**, *91*. [CrossRef]

22.   Liu, Q.; Guan, W.H.; Long, S.B.; Jia, R.; Liu, M.; Chen, J.N. Resistive switching memory effect of ZrO2 films with Zr$^+$ implanted. *Appl. Phys. Lett.* **2008**, *92*. [CrossRef]

23.   Chen, J.-Y.; Hsin, C.-L.; Huang, C.-W.; Chiu, C.-H.; Huang, Y.-T.; Lin, S.-J.; Wu, W.-W.; Chen, L.-J. Dynamic evolution of conducting nanofilament in resistive switching memories. *Nano Lett.* **2013**, *13*, 3671–3677. [CrossRef] [PubMed]

24.   Wedig, A.; Luebben, M.; Cho, D.-Y.; Moors, M.; Skaja, K.; Rana, V.; Hasegawa, T.; Adepalli, K.K.; Yildiz, B.; Waser, R.; *et al.* Nanoscale cation motion in TaO$_x$, HfO$_x$ and TiO$_x$ memristive systems. *Nat. Nanotechnol.* **2016**, *11*. [CrossRef] [PubMed]

25.   Yu, S.; Chen, H.-Y.; Gao, B.; Kang, J.; Wong, H.S.P. HfO$_x$-based vertical resistive switching random access memory suitable for bit-cost-effective three-dimensional cross-point architecture. *ACS Nano* **2013**, *7*, 2320–2325. [CrossRef] [PubMed]

26. Kim, K.-H.; Gaba, S.; Wheeler, D.; Cruz-Albrecht, J.M.; Hussain, T.; Srinivasa, N.; Lu, W. A functional hybrid memristor crossbar-array/CMOS system for data storage and neuromorphic applications. *Nano Lett.* **2012**, *12*, 389–395. [CrossRef] [PubMed]

27. Chasin, A.; Zhang, L.Q.; Bhoolokam, A.; Nag, M.; Steudel, S.; Govoreanu, B.; Gielen, G.; Heremans, P. High-performance a-IGZO thin film diode as selector for cross-point memory application. *IEEE Electron Device Lett.* **2014**, *35*, 642–644.

28. Li, Y.T.; Lv, H.B.; Liu, Q.; Long, S.B.; Wang, M.; Xie, H.W.; Zhang, K.W.; Huo, Z.L.; Liu, M. Bipolar one diode-one resistor integration for high-density resistive memory applications. *Nanoscale* **2013**, *5*, 4785–4789. [CrossRef] [PubMed]

29. Park, J.B.; Biju, K.P.; Jung, S.; Lee, W.; Lee, J.; Kim, S.; Park, S.; Shin, J.; Hwang, H. Multibit operation of TiO_x-based ReRAM by Schottky barrier height engineering. *IEEE Electron Device Lett.* **2011**, *32*, 476–478. [CrossRef]

30. Chu, T.J.; Tsai, T.M.; Chang, T.C.; Chang, K.C.; Pan, C.H.; Chen, K.H.; Chen, J.H.; Chen, H.L.; Huang, H.C.; Shih, C.C.; *et al.* Ultra-high resistive switching mechanism induced by oxygen ion accumulation on nitrogen-doped resistive random access memory. *Appl. Phys. Lett.* **2014**, *105*. [CrossRef]

31. Vanheusden, K.; Warren, W.L.; Seager, C.H.; Tallant, D.R.; Voigt, J.A.; Gnade, B.E. Mechanisms behind green photoluminescence in ZnO phosphor powders. *J. Appl. Phys.* **1996**, *79*, 7983–7990. [CrossRef]

32. Celano, U.; Goux, L.; Degraeve, R.; Fantini, A.; Richard, O.; Bender, H.; Jurczak, M.; Vandervorst, W. Imageing the three-dimensional conductive channel in filamentary-based oxide resistive switching memory. *Nano Lett.* **2015**, *15*. [CrossRef] [PubMed]

33. Xie, H.W.; Liu, Q.; Li, Y.T.; Lv, H.B.; Wang, M.; Liu, X.Y.; Sun, H.T.; Yang, X.Y.; Long, S.B.; Liu, S.; *et al.* Nitrogen-induced improvement of resistive switching uniformity in a HfO2-based RRAM device. *Semicond. Sci. Technol.* **2012**, *27*. [CrossRef]

34. Allen, M.W.; Durbin, S.M. Influence of oxygen vacancies on Schottky contacts to ZnO. *Appl. Phys. Lett.* **2008**, *92*. [CrossRef]

![nanomaterials logo] *nanomaterials*

MDPI

Article

Nanostructuring of Palladium with Low-Temperature Helium Plasma

P. Fiflis *, M.P. Christenson, N. Connolly and D.N. Ruzic

Department of Nuclear, Plasma and Radiological Engineering, Center for Plasma Material Interactions, University Illinois at Urbana-Champaign, Urbana 61801, IL, USA; mpchris2@illinois.edu (M.P.C.); connlly2@illinois.edu (N.C.); druzic@illinois.edu (D.N.R.)

* Author to whom correspondence should be addressed; fiflis1@illinois.edu; Tel.: +1-708-655-1432.

Academic Editors: Krasimir Vasilev and Thomas Nann

Received: 13 October 2015; Accepted: 20 November 2015; Published: 25 November 2015

Abstract: Impingement of high fluxes of helium ions upon metals at elevated temperatures has given rise to the growth of nanostructured layers on the surface of several metals, such as tungsten and molybdenum. These nanostructured layers grow from the bulk material and have greatly increased surface area over that of a not nanostructured surface. They are also superior to deposited nanostructures due to a lack of worries over adhesion and differences in material properties. Several palladium samples of varying thickness were biased and exposed to a helium helicon plasma. The nanostructures were characterized as a function of the thickness of the palladium layer and of temperature. Bubbles of ~100 nm in diameter appear to be integral to the nanostructuring process. Nanostructured palladium is also shown to have better catalytic activity than not nanostructured palladium.

Keywords: palladium; nanotendrils; helium bubbles; catalysis

1. Introduction

Experiments at several different institutions have observed the growth of tungsten nanostructures under exposure to helium plasmas while investigating the viability of tungsten for high heat flux components in nuclear fusion reactors [1,2]. While these structures are potentially fatal to the fusion plasma when grown [3], they do exhibit characteristics that could be exploited in other applications. A high porosity, a low density of about 10% of the bulk material, large surface area, increased emissivity, and decreased reflectance are all properties of the nanostructured surface [4–7]. The nanostructures are produced by prolonged exposure to a flux of helium ions while the tungsten is at an elevated temperature. Several studies have investigated the similar formation of nanostructures on metals other than tungsten. Exposure to fluxes of helium in excess of 10^{20} m^{-2}·s^{-1} at a temperature approximately between 30% and 50% of the melting temperature yields several different nanostructures, *i.e.*, cones/pillars have been observed on copper [8], fuzz on molybdenum and tungsten [5], and roughening of the surface on titanium [8]. Palladium is widely used as a catalyst in a variety of different chemical reactions; and because catalytic activity scales linearly with the surface area of the catalyst, experiments were undertaken to investigate the nanostructuring of palladium under irradiation by helium ions. To this end, several palladium substrates were exposed at elevated temperatures to helium ions from a helicon plasma source. The dependence of the formed nanostructures on temperature and sample geometry are described herein.

2. Experimental Section

Palladium samples (Alfa Aesar 99.9%, Ward Hill, MA, USA) were exposed inside of a commercial grade helium helicon source (MORI 200, Trikon Technologies, Newport, UK) [9]). The plasma conditions for the experiments described herein were generated with an RF power of 700 W, a magnetic field of 120 G, and a background helium pressure of 100 mTorr as read by a convectron gauge (Granville Phillips 375, MKS Instruments, Andover, MA, USA). A photo of the experimental chamber can be seen in Figure 1. The resulting plasma density is 1×10^{18} m^{-3} with an electron temperature of 4 eV diagnosed with an RF-compensated Langmuir probe [10] in the region where the sample was placed. The palladium was supported via a copper sample holder, which suspended the sample in the plasma. The sample was biased to negative 40 V with respect to plasma potential such that the incoming helium ion flux had an energy of 40 eV and a flux of 2.5×10^{21} m$^{-2} \cdot$s^{-1}. Sample temperature was achieved merely by heating of the sample via the incoming ion flux. Regulation of the temperature, however, was achieved by adjusting the area of the sample in direct contact with the copper sample holder, thereby controlling conduction losses. It should be noted that the centerline density of the plasma is constant to within measurement error over the range of sample placements, so adjustment of the sample relative to the copper sample holder did not change the flux to the target. Temperatures were not directly measured, but rather computed via an experimentally calibrated finite difference model which balances input energy from helium ion irradiation and losses via conduction and radiation [11]. Scanning electron microscopy (Hitachi S4700, Tokyo, Japan) was performed on the exposed samples. Four different geometries were tested; a plate of palladium 1 cm × 2 cm × 0.5 mm, a wire 0.5 mm diameter × 20 cm in length, and two thin films deposited on 25 mm × 25 mm glass substrates with thicknesses of 300 nm (evaporation coating) and 30 nm (magnetron sputter coating).

Figure 1. Photo of exposure chamber showing MORI (Trikon Technologies, Newport, UK) automated matching network, exposure volume, load lock gate valve, and transfer arm for introducing samples without breaking vacuum.

3. Results and Discussion

3.1. Palladium Nanostructuring as a Function of Temperature

Scanning electron microscope (SEM) micrographs of the exposed wire and plate are shown in Figures 2–4, respectively. Figure 3 is a series of top down (0° from normal) micrographs of the plate; Figure 4 is a series of micrographs of the plate and has a tilt of 40° from normal introduced to the sample. From these micrographs it can be seen that exposure of palladium at elevated temperatures to fluxes of helium ions forms a series of pillars or tendrils from the surface. Energy dispersive X-ray Spectroscopy (EDX) analysis of these tendrils confirms that they are palladium. These tendrils are 350 ± 100 nm in diameter, 1000 ± 250 nm in height, and possess an areal density of approximately 1.5 tendrils/μm^2. It can also be seen from the difference between the 0° and 40° micrographs of the plate that these tendrils grow normally to the surface. The tendrils are also mostly straight at temperatures less than 700 K, and in excess of 700 K the structures begin to bend and fold. As the temperature increases beyond this (>775 K), the nanostructures begin to increase in diameter (500 ± 100 nm), but not length. The areal density of tendrils decreases to approximately 1 tendril/μm^2. As the temperature exceeds 850 K, the nanostructures appear to begin to anneal back into the bulk, resulting in very rounded and thick tendrils at 880 K (see Figure 2), and full disappearance of tendrils by 900 K (Figure 5). The nanostructuring process appears to be bubble driven, similar to the growth of tungsten nanostructured "fuzz" which is predicted to grow under certain conditions in the divertor region of fusion reactors [12]. However, while the bubbles that drive nanostructuring in tungsten are approximately 10 nm in diameter, the pits observed in the surface of the nanostructured palladium (attributed to bubble bursting at the surface similar to tungsten [11]) are 75 ± 25 nm in diameter at temperatures less than 750 K. At temperatures above 750 K, these pits swell in size to 110 ± 30 nm in diameter. Normalizing two characteristic parameters of the nanostructures to the pit diameter draws striking parallels between tungsten and palladium nanostructuring. The ratio of tendril diameter to pit diameter in tungsten is approximately 3 to 4 [11]. Similarly, the ratio of tendril diameter to pit diameter in palladium is also 3 to 4. Additionally, the separation distance between individual tendrils is approximately 7 to 11 times the diameter of the pits in tungsten [11]. With a tendril separation distance of 800 ± 150 nm, palladium nanostructures have a ratio of tendril separation distance to pit diameter in the exact same range. This is very indicative of a similar bubble mechanism driving the growth of fuzz tendrils, whereby bubbles created in the bulk of the material effectively rise to the surface or grow and thin the material above them, subsequently rupturing. As more bubbles impact the surface, hills and valleys start to form stochastically. Bubbles then are more likely to connect to a valley rather than a hill by virtue of shorter path length, and as a result nanostructures grow. Characteristics such as the ratio of tendril diameter or tendril separation distance to the pit diameter fall out of a simple Monte Carlo model implementing only the assumption that bubbles are more likely to rupture at a valley than a hill [11]. Since the model is independent of material properties, any nanostructuring via the same mechanism will show similar tendril to pit diameter ratios and tendril separation to pit diameter ratios, as is seen here. Much like tungsten nanostructuring via helium plasma bombardment, palladium nanostructuring appears to have a window of temperature for which it can grow tendrils. Tungsten fuzz grows within the temperature range of 1000–2000 K (0.27–0.54 T_m) [13]. Palladium nanostructuring in the experiments described herein occurred at 650 K (0.33 T_m) and 880 K (0.48 T_m) as well as several other intermediate temperatures and was bracketed by a lack of nanostructuring formation at 500 K (0.27 T_m) and 900 K (0.49 T_m). When normalized to the melting point of the material, the active temperature range for the bubble driven nanostructuring appears to be very similar.

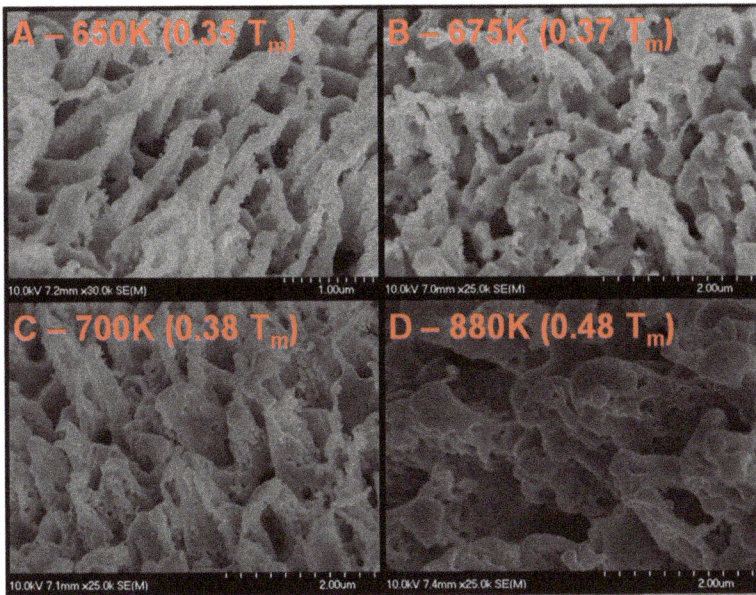

Figure 2. Scanning electron microscope (SEM) micrographs of palladium surface (0.5 mm diameter wire sample) after exposure to helium plasma at elevated temperature. The flux to each area is identical, the only changed variable is temperature (noted in the upper left corner of each micrograph both absolute and as a fraction of the melting point of palladium).

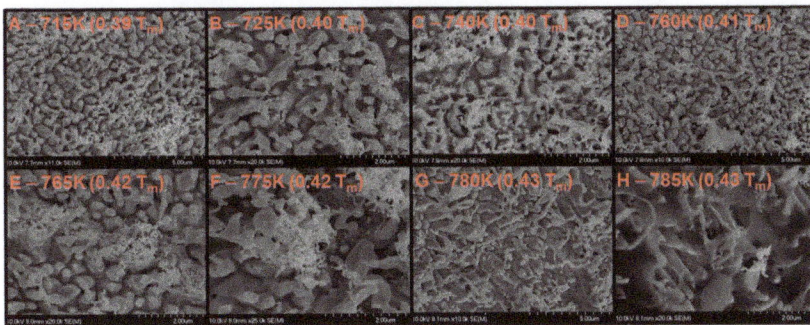

Figure 3. SEM micrographs of palladium surface (0.5 mm plate sample) after exposure to helium plasma at elevated temperature. The flux to each area is identical, the only changed variable is temperature (noted in the upper left corner of each micrograph both absolute and as a fraction of the melting point of palladium). Secondary electron collection performed at a tilt angle of $0°$ with respect to the surface normal.

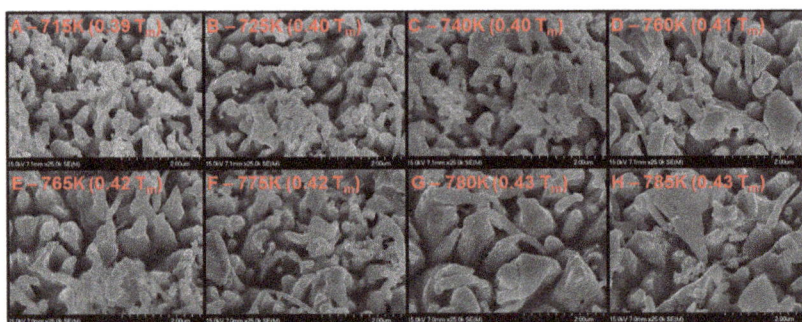

Figure 4. SEM micrographs of palladium surface (0.5 mm plate sample) after exposure to helium plasma at elevated temperature. The flux to each area is identical, the only changed variable is temperature (noted in the upper left corner of each micrograph both absolute and as a fraction of the melting point of palladium). Secondary electron collection performed at a tilt angle of 40° with respect to the surface normal.

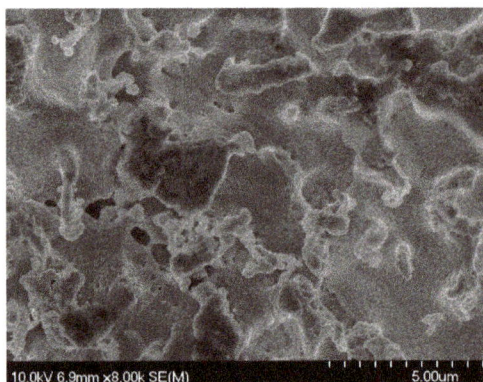

Figure 5. SEM micrograph of palladium surface (0.5 mm diameter wire sample) after exposure to helium plasma at 900 K, only a couple tendrils are visible as the annealing rate of the tendrils begins to exceed the rate of growth.

3.2. Palladium Nanostructuring versus Palladium Thickness

Many applications of palladium nanostructuring revolve around the increased surface area to volume ratio (*i.e.*, improving the catalytic activity for a given weight of palladium). As a result of this, the ability to grow nanostructures on thin films of palladium is very desirable as this would further increase the surface area to volume ratio of an amount of palladium with an already large surface area to volume ratio. There, however, was thought to be a minimum thickness at which the palladium would no longer nanostructure because the bubbles necessary to drive the growth of palladium nanostructures were approximately 100 nm in diameter and could therefore not grow to full size in very thin films. As thick substrates of 0.5 mm thickness and diameter had already been tested, two thin films (of thickness 300 and 30 nm) were investigated to see the structures that would form. SEM micrographs of each sample can be seen in Figures 6 and 7, respectively. Nanostructure growth for the 300 nm film appears to be again bubble driven. Tendrils of diameter 320 ± 100 nm are apparent and so are pits of 75 ± 15 nm diameter. These are commensurate with the diameters of the tendrils and pits of the bulk samples at the same temperature of exposure. The tendrils are also very straight, much like the tendrils observed in the bulk samples at temperatures <700 K. This implies the bubble formation depths and subsequent loop punching to the surface occurs at depths of less than 3 bubble diameters, much like the formation of helium bubbles within tungsten which drive nanostructuring [11]. As the thickness of the film is reduced below that of the bubble diameter, formation of full helium bubbles to drive nanostructuring is suppressed. Instead, it appears as though formation and growth of bubbles within the 30 nm thick film rupture the film without being able to build upon each other and grow nanostructures. This results in a series of pits in the surface, but no vertical growth of nanostructures. These pits are of diameter 130 ± 35 nm, which is larger than those observed in the 300 nm and bulk samples. Wrinkles are also evident in the palladium film which is indicative of delamination of the palladium film from the SiO_2 substrate. The palladium film was deposited via magnetron sputtering at room temperature and due to residual tensile stresses in the film, once it became delaminated, it wrinkled.

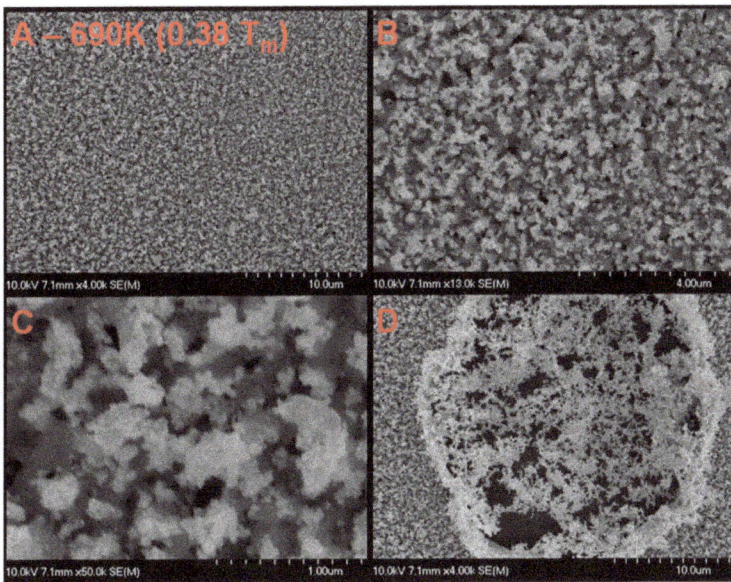

Figure 6. SEM micrograph of palladium surface (300 nm thin film deposited on SiO_2) after exposure to helium plasma at elevated temperature. (**A**), (**B**), and (**C**) are different resolutions of the same location showing growth of tendrils and voids that appear to penetrate down to the SiO_2 substrate. Tendrils approximately the same diameter as those observed on bulk Pd samples are observed. Pits of similar diameter are also observed. (**D**) shows an area of the palladium film where the helium plasma has eroded through the palladium film to the substrate with very thin tendrils of Pd stretching across.

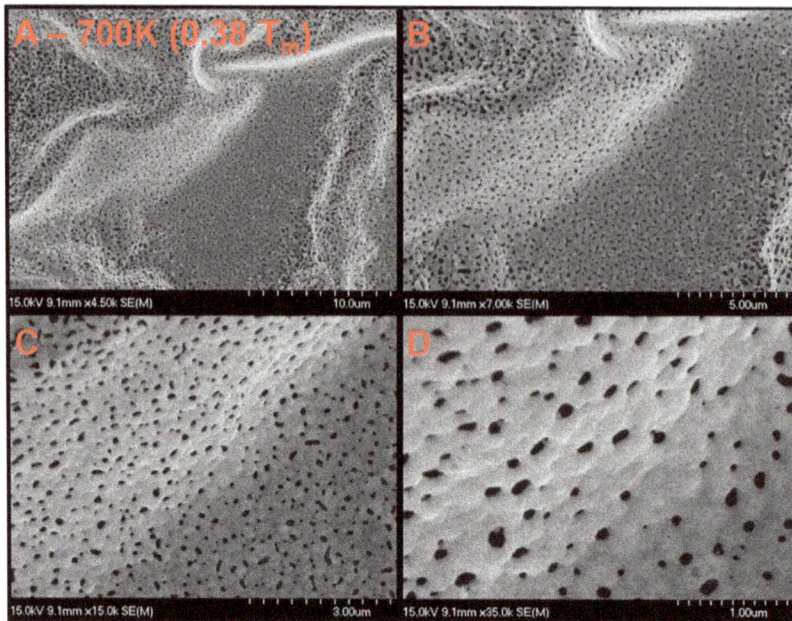

Figure 7. SEM micrograph of palladium surface (30 nm thin film deposited on SiO₂) after exposure to helium plasma at elevated temperature. Figure 5A–D are different resolutions of the same location showing growth no tendril growth, but a significant amount of voids. These voids are of a diameter greater than the pits observed in the bulk and 300 nm film samples. Large wrinkles appear evident in the film. It appears as though formation and growth of bubbles within the 30 nm thick film rupture the film without being able to build upon each other and grow nanostructures.

3.3. Catalysis with Nanostructured Palladium

It has been suggested that the increased surface area of nanostructured palladium could provide greatly enhanced catalytic activity. To investigate this hypothesis, a comparison was made between identical plates, one not nanostructured and the other nanostructured under identical conditions to the plate described above. This comparison of catalytic properties was carried out using a reduction reaction, which modified cyclohexene into cyclohexane through the syn addition of two hydrogens using the palladium pieces as the catalysts for the reaction [14]:

$$C_6H_{10} + H_2 \rightarrow C_6H_{12}$$

The catalytic properties of each of the palladium samples, both smooth and nanostructured, were compared to the industrial palladium catalytic standard of palladium absorbed onto a carbon surface [15]. This standard is widely accepted as the best way to increase surface area for catalytic reactivity.

Alkenes are reduced to alkanes through a multi-body process, where the palladium or other catalyst acts as an activation site for the reaction. Following the Horiuti-Polanyi mechanism [14], hydrogen first dissociatively chemisorbs to the bonding site on the palladium surface, but only if the hydrogen molecule has its axis parallel to the surface of the palladium crystal [16]. For this reason, hydrogen is often found in excess in these reactions. Cyclohexene, present in the reaction in liquid form, dissociatively bonds to the surface of the palladium in much the same way as the hydrogen molecule, where the pi bond in the double bond between the carbons is broken. The carbons then bond to the palladium surface. The dissociated hydrogen atoms then bond to the free sites on the carbon atoms currently bonded to the palladium surface and the molecule leaves the palladium surface as an alkane since alkanes are not strongly adsorbed on the surface. For a single reaction on a simple palladium surface, this leads to a significant reduction in the activation energy, making the reaction feasible for scale-up.

All reactions were carried out with the primary reactant, cyclohexene, in liquid phase. A block diagram of the experimental setup can be seen in Figure 8. The reaction vessel was a 500 mL, three-neck round bottom flask connected to a vacuum line on one neck, a gas bubbler on the second neck, and stoppered on the third neck. Prior to closing off the flask, the catalytic samples were weighed and added, along with a magnetic stir rod. The reaction chamber was then sealed. The chamber was then purged with argon and evacuated three separate times to ensure atmospheric purity. 150 mL of cyclohexene was added through the stopper using a syringe. A second syringe connected to a hydrogen line was then inserted into the cyclohexene liquid in order to bubble into the flask an excess of hydrogen gas, as noted by the bubbler. Before hydrogen was added, a control sample was taken. Samples were then taken using a microsyringe at 30 min intervals for 180 min, while the hydrogen was bubbled through the reactants and the stir rod agitated the reactants. This was done to test the reaction rate for each of the catalyst types. All samples taken were 100 µL in volume and were added to NMR sample tubes along with 600 µL of deuterated chloroform ($CDCl_3$), which were mixed through inversion.

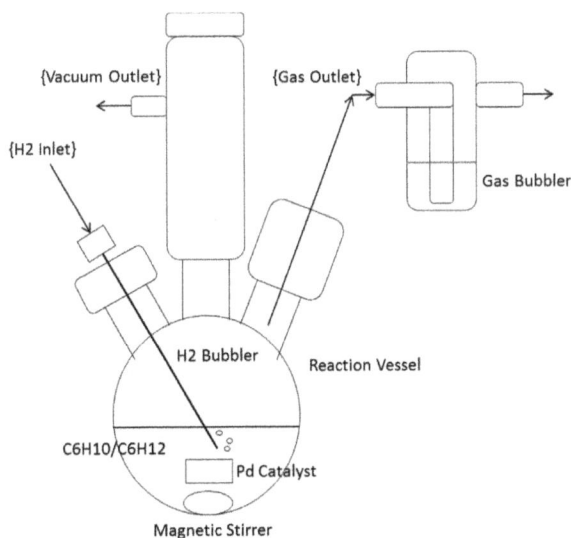

Figure 8. A block diagram of the palladium-catalyzed hydrogenation reaction vessel, including the hydrogen gas inlet, the vacuum cylinder outlet, the magnetic stirrer, and the gas bubbler used to qualitatively determine the flow rate of the hydrogen through the reactant volume.

The catalytic capability of the different palladium types was measured by determining the ratios of the areas under the typical cyclohexene peaks to the typical cyclohexane peaks seen in Nuclear magnetic resonance (NMR) spectra. A Varian Unity Inova NMR spectroscopy machine [17,18] was used to make the measurements, and each scan used a spectral frequency of 399.74 MHz, with 1 transient at an acquisition time of 4.096 seconds. The scans were performed using one-dimensional proton NMR at a 45 degree pulse width. Ratios were taken between areas under the 1.4 ppm cyclohexane peak and the mean of the areas under the 1.2 ppm, 1.7 ppm, and 2.0 ppm cyclohexene peaks, normalized to the deuterated chloroform standard. Figure 9 shows the results of the catalytic conversion over time for the different catalyst types, indicating that the fuzzed sample is greatly improved in conversion over the non-fuzzed sample.

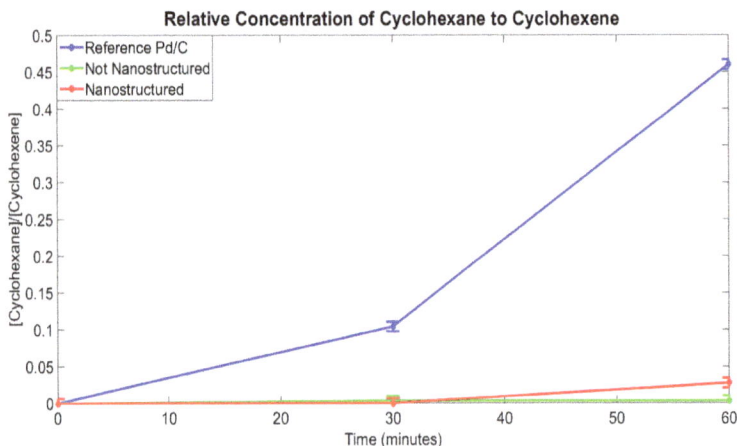

Figure 9. A plot of the yield measured as the ratio of the concentration of the cylcohexane to cyclohexene. This ratio is based on the ratios of areas under the curve of the cyclohexane 1.4 ppm peak and the mean of the intensities of the three cyclohexene peaks seen at 1.2 ppm, 1.7 ppm, and 2.0 ppm in the NMR scans, which are characteristic for the respective compounds and are normalized to the deuterated chloroform standard. These ratios were also taken as a function of time to, not only observe the effectiveness of each catalyst type, but also how the kinetics of the reaction compare with each catalyst type.

4. Discussion and Conclusions

A variety of palladium samples were exposed to a flux of 40 eV helium ions at temperatures between 0.3 and 0.5 T_m. Samples of bulk palladium (*i.e.*, wire and plate) showed evidence of bubbles of approximate diameter 100 nm and tendrils of approximate diameter 350 nm. The nanostructuring growth mechanism appears to be similar to that of tungsten, with an active temperature range similar to that of tungsten after normalization to the melting point. However, the diameter of the bubbles is much larger than that of those observed in tungsten. Previous studies of exposure of different metals to energetic helium fluxes at elevated temperatures have suggested that the nanostructuring process is heavily dependent on crystal structure [8]. Body centered cubic (bcc) crystals, such as tungsten, molybdenum, and tantalum, show very similar nanostructures in both size and morphology. Palladium is a face centered cubic (fcc) material, and therefore, will nanostructure differently than the bcc tungsten. However, since the ratio of tendril diameter to pit diameter as well as the ratio of inter-tendril separation to pit diameter is the same for both tungsten and palladium, it is highly probable that the mechanism for the formation of the nanostructures is the same. The difference in bubble size then is the biggest driver in the difference in observed morphology.

Nanostructuring the palladium plate resulted in a large increase in catalytic activity beyond that of the non-nanostructured sample. It should, however, be noted that the reference Pd/C catalyst outperformed both of the fuzzed and non-fuzzed samples due to the very large surface area provided by the activated carbon. Mechanical removal of these nanostructures from the surface of the palladium to produce a finely nanostructured powder catalyst may increase the catalytic activity beyond that of the standard. Alternatively, nanostructuring of other geometries with a helium plasma may offer advantages in systems where filtration of powder catalysts would be impractical and conventional methods of surface roughening, such as sand blasting, would be too violent on fragile catalyst geometries. Nanostructuring by helium plasma could also be performed atop a sand blasted layer to further increase the surface area.

Acknowledgments: The authors would like to Samuel Yick (CSIRO, Australia) for preparation of the 30 nm thin film palladium sample, as well as Lynford Goddard and Steven McKeown for preparation of the 300 nm palladium sample, and Dean Olson for running NMR tests and providing guidance in NMR analyzation.

Author Contributions: The nanostructuring experiments were conceived and designed by P. Fiflis and D.N. Ruzic. P. Fiflis carried out the nanostructuring experiments. Chemical analysis was designed and performed by M.P. Christenson and N. Connolly. Data analysis performed by P. Fiflis, M.P. Christenson, and N. Connolly. All authors contributed to the composition of the paper.

Conflicts of Interest: The authors declare no conflict of interest.

References

1. Wright, G.M.; Brunner, D.; Baldwin, M.J.; Bystrov, K.; Doerner, R.P.; Labombard, B.; Lipschultz, B.; De Temmerman, G.; Terry, J.L.; Whyte, D.G.; *et al.* Comparison of tungsten nano-tendrils grown in Alcator C-Mod and linear plasma devices. *J. Nuc. Mat.* **2013**, *438*, S84–S89. [CrossRef]
2. Baldwin, M.J.; Doerner, R.P. Formation of helium induced nanostructure "fuzz" on various tungsten grains. *J. Nuc. Mat.* **2010**, *404*, 165–173. [CrossRef]
3. Pitts, R.A.; Carpentier, S.; Escourbiac, F.; Hirai, T.; Komarov, V.; Lisgo, S.; Kukushkin, A.S.; Loarte, A.; Merola, M.; Sashala Naik, A.; *et al.* A full tungsten divertor for ITER: Physics issues and design status. *J. Nuc. Mat.* **2013**, *438*, S48–S56. [CrossRef]
4. Takamura, S.; Ohno, N.; Nishijima, D.; Kajita, S. Formation of nanostructured tungsten with arborescent shape due to helium plasma irradiation. *Plas. Fus. Res.* **2006**, *1*. [CrossRef]
5. De Temmerman, G.; Bystrov, K.; Zielinski, J.J.; Balden, M.; Matern, G.; Arnas, C.; Marot, L. Nanostructuring of molybdenum and tungsten surfaces by low-energy helium ions. *J. Vac. Sci. Technol. A* **2012**, *30*, 041306. [CrossRef]
6. Baldwin, M.J.; Doerner, R.P. Helium induced nanoscopic morphology on tungsten under fusion relevant plasma conditions. *Nucl. Fus.* **2008**, *48*, 3. [CrossRef]
7. Kajita, S.; Ono, N.; Yokochi, T.; Yoshida, N.; Yoshihara, R.; Takamura, S.; Hatae, T. Optical properties of nanostructured tungsten in near infrared range. *Plas. Phys. Contr. Fus.* **2012**, *54*, 10. [CrossRef]
8. Tanyeli, I.; Marot, L.; Mathys, D.; van de Sanden, M.C.M.; De Temmerman, G. Surface modifications induced by high fluxes of low energy helium ions. *Sci. Reports.* **2015**, *5*. [CrossRef] [PubMed]
9. Tobe, R.; Sekiguchi, A.; Sasaki, M.; Okada, O.; Hosokawa, N. Plasma-enhanced CVD of TiN and Ti using low-pressure and high-density helicon plasma. *Thin Solid Films* **1996**, *281–282*, 155–158. [CrossRef]
10. Ruzic, D.N. *Electric Probes for Low Temperature Plasmas*, 1st ed.; American Vacuum Society Education Committee: New York, NY, USA, 1994.
11. Fiflis, P.; Curreli, D.; Ruzic, D.N. Direct Time-Resolved Observation of Tungsten Nanostructured Growth Due to Helium Plasma Exposure. *Nucl. Fus.* **2015**, *55*, 3. [CrossRef]
12. Brooks, J.N.; Allain, J.P.; Doerner, R.P.; Hassanein, A.; Nygren, R.; Rognlien, T.D.; Whyte, D.G. Plasma–surface interaction issues of an all-metal ITER. *Nucl. Fus.* **2009**, *49*, 035007. [CrossRef]
13. Ueda, Y.; Peng, H.Y.; Lee, H.T.; Ohno, N.; Kajita, S.; Yoshida, N.; Doerner, R.; De Temmerman, G.; Alimov, V.; Wright, G. Helium effects on tungsten surface morphology and deuterium retention. *J. Nuc. Matl.* **2013**, *442*, S267–S272. [CrossRef]
14. Horiuti, I.; Polanyi, M. Exchange Reactions of Hydrogen on Metallic Catalysts. *T. Faraday Soc.* **1934**, *30*, 1164–1172. [CrossRef]
15. Puskás, R.; Sápi, A.; Kukovecz, A.; Kónya, Z. Comparison of Nanoscaled Palladium Catalysts Supported on Various Carbon Allotropes. *Top. Catal.* **2012**, *55*, 865–872. [CrossRef]
16. Lischka, M.; Groß, A. Hydrogen on palladium: A model system for the interaction of atoms and molecules with metal surfaces. *Recent Dev. Vac. Sci. Technol.* **2003**, *37*, 111–132.

17. McKenna, S.; Spyracopoulos, L.; Moraes, T.; Pastushok, L.; Ptak, C.; Xiao, W.; Ellison, M.J. Noncovalent Interaction between Ubiquitin and the Human DNA Repair Protein Mms2 Is Required for Ubc13-mediated Polyubiquitination. *J. Bio. Chem.* **2001**, *276*, 40120–40126. [CrossRef] [PubMed]
18. Miura, K.; Ohgiya, S.; Hoshino, T.; Nemoto, N.; Suetake, T.; Miura, A.; Spracopoulos, L.; Kondo, H.; Tsuda, S. NMR Analysis of Type III Antifreeze Protein Intramolecular Dimer. *J. Bio. Chem.* **2001**, *276*, 1304–1310. [CrossRef] [PubMed]

nanomaterials

MDPI

Review

Selective Plasma Etching of Polymeric Substrates for Advanced Applications

Harinarayanan Puliyalil [1,2,†] and Uroš Cvelbar [1,2,*,†]

1   Jožef Stefan International Postgraduate School, Jamova cesta 39, 1000 Ljubljana, Slovenia; hari.puliyalil@ijs.si
2   Jožef Stefan Institute, Jamova cesta 39, 1000 Ljubljana, Slovenia
*   Correspondence: uros.cvelbar@ijs.si; Tel.: +386-14773536
†   These authors contributed equally to this work.

Academic Editors: Krasimir Vasilev and Melanie Ramiasa
Received: 2 February 2016; Accepted: 30 May 2016; Published: 7 June 2016

Abstract: In today's nanoworld, there is a strong need to manipulate and process materials on an atom-by-atom scale with new tools such as reactive plasma, which in some states enables high selectivity of interaction between plasma species and materials. These interactions first involve preferential interactions with precise bonds in materials and later cause etching. This typically occurs based on material stability, which leads to preferential etching of one material over other. This process is especially interesting for polymeric substrates with increasing complexity and a "zoo" of bonds, which are used in numerous applications. In this comprehensive summary, we encompass the complete selective etching of polymers and polymer matrix micro-/nanocomposites with plasma and unravel the mechanisms behind the scenes, which ultimately leads to the enhancement of surface properties and device performance.

Keywords: plasma etching; selectivity; surface chemistry; nanomesh; nanostructuring; polymer composite; biomaterials; adhesion; etch mask; radical atoms

---

1. Introduction

Plasma technology is one of the fastest developing branches of science, which is replacing numerous conventional wet-chemical methods in high-tech laboratories and industries, with a huge impact in renewable energy, environmental protection, biomedical applications, nanotechnology, microelectronics, and other fields. Plasma, the complex mixture of ions, radicals, electrons, and excited molecules has replaced conventional methods to develop various nanostructured materials with complex morphology and advanced properties (e.g., the production of vertically aligned carbon nanotubes (CNTs), which is difficult to achieve with other synthetic methods [1,2]). Owing to fast multi-scale production, it is a preferred method for the synthesis of other nanomaterials such as nanowires (NWs), carbon nanowalls (CNW), graphene sheets, *etc.* [3–5]. Through this, plasma in nanoscience also impacts the field of renewable energy and environmental protection, where we endeavor to replace hydrocarbon energy resources with solar energy or hydrogen fuel cells [6,7]. An additional benefit of plasma technology is the wet free doping of semi-conducting nanomaterials with heteroatoms to alter the band gap energy and conductivity for various applications [8–10]. In the overview of the applications in biology and medicine, plasma is found in various applications, from plasma surgery and the manufacture of artificial implants to the straightforward disinfection of the medical equipment.

The main processing routes in plasma technology are simply concluded into deposition and etching. For achieving the best material performance, these two techniques are either used individually or in a combination. Plasma-enhanced chemical vapor deposition (PECVD) is a technique complementary to thermally excited chemical vapor deposition (CVD) and is used for creating thin layer coatings, micro- or nanostructures, and even the deposition of complex functional materials [11]. Various applications of plasma deposition techniques are found in numerous articles, and some examples are given in [12–15]. On the other hand, plasma etching originates from the interactions between plasma particles with various substrates. The interactions are either physical or chemical; the particles with high kinetic energy are utilized to knock out the atomic or molecular species from the surface, whereas in the latter case the interaction of particles is purely potential. The physical interactions are mostly linked to ionized plasma species, while potential is linked to neutral or excited species.

Plasma etching, referred many times as plasma chemical etching or dry etching, of both organic and inorganic materials was reported for material fabrication in multidisciplinary applications. Compared to the wet chemical etching, plasma etching is capable of controlled and precise etching at very small scales (~10 nm). Additionally, plasma etching limits the disadvantages, for example, via contamination or solvent absorption during the treatment process. Some of the major advantages and disadvantageous of wet chemical etching *versus* plasma etching are listed in Table 1. In this review, we will mostly focus on the plasma etching of organic materials, which typically undergo etching relatively faster than inorganic materials. The difference in the etching arises from the stability of materials towards various chemical species. As an example, when we compare with liquid chemistry, alkali metals react explosively with water or acids, whereas transition or inner transition metals react very slowly or even stay inert. In the same way, different organic or inorganic materials react at different rates with various plasma species. This advantage of the disparate reactivity of substrates towards plasma is then utilized for fabricating important micro-/nanostructured materials.

Table 1. Comparison of wet chemical etching *versus* plasma etching.

	Wet Chemical Etching	Plasma Etching
Etchant:	Chemical (acids, alkali, *etc.*).	Reactive gas (radicals, ions, *etc.*).
Etch rate and selectivity:	High.	Good, controllable.
Advantageous:	Low equipment cost, fast processing and easy to implement.	Capable of small scale etching (~10 nm), no contamination issues, no hazardous chemicals, ecologically benign technology.
Disadvantageous:	Inadequate to define small feature size less than 1 µm, handling of hazardous chemicals, contamination issues, ecologically unfriendly technology with need of waste processing.	High equipment cost, implementation dependent on application, potential radiation damage.
Directionality:	Only isotropic etching.	Can be isotropic or anisotropic.

In the field of material fabrication, the organic materials, especially polymer materials, are replacing many inorganic substrates. Simple fabrication and low production costs are the two great advantages for employing polymer materials. However, to minimize certain disadvantages of pure polymers, their properties are improved by making composites from multiple organic materials using multiplication of material properties or even reducing their dimensions to the nanoscale. To achieve this, the preferential removal of materials is desired, which is simply achieved by plasma etching. This approach was found to be the best method for many applications to achieve desired surface structures, morphology, or chemistry [16,17]. One should stress the effects on the surface chemistry, especially when it comes to organic materials, as it is typically more important than the morphology for certain applications such as adhesion. Plasma etching always alters the surface energy by the successful incorporation or formation of chemical groups during the interaction of plasma particles with materials [18].

Up to now, no reviews have reported specifically on the topic of plasma selective etching, which has innumerable applications in the field of materials science. This review concentrates on the applications of selective etching. However, prior to that, a short discussion is made on the important types of plasmas based on particle species, their energy, and the result of interaction with the surface, namely sputtering, reactive ion etching, or neutral radical chemical etching. All these interactions then drive the representative applications of plasma etching schematically presented in Figure 1. Sputtering, which involves the removal of various substrates merely by high-energy particle collision is not very important for the selective etching of materials. On the contrary, the reactive ion etching and the neutral plasma chemical etching are found to be efficient for the preferential removal of one material over the other. Following this, the applications of the plasma modified polymeric substrates are described. The main objective of this article is to focus on the characteristics of polymer materials, which are decisive for the etching rate and stability towards high-energy particles. Plasma processing of various polymer composites and their advanced applications are outlined as well.

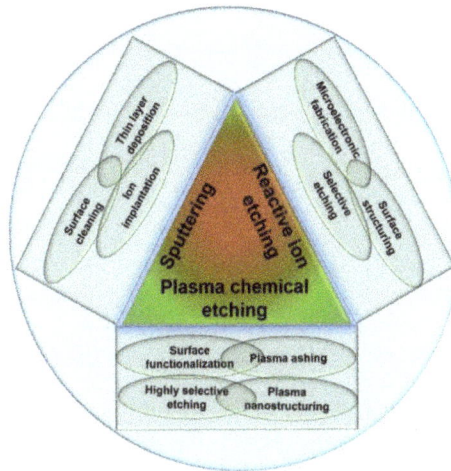

Figure 1. Schematic representation of various uses of different plasma processes.

2. Plasma Processing

2.1. Sputtering

Sputtering deposition is widely employed as a robust method in a thin film preparation for countless applications. The basic principle of the sputtering technique is the generation of high-energy ions, which are then accelerated and guided towards the substrate where they knock out surface atoms as ions into the vapor phase. The so-generated ions are also guided and deposited on the coated substrate where they form a thin layer. However, this deposition process is beyond the scope of this paper. More interesting are energetic surface collisions, where material is removed from the surface.

While considering the process, one should create an ion-rich environment that enables the acceleration of generated particles by applying an external electric or magnetic field. The number density and kinetic energy of the ions (N_{ion} and $W_{K.E.}$, respectively) should be sufficiently large to overcome the energy barrier created by the binding energy of the surface atoms ($W_{B.E.}$). The sputtering process is efficient when the kinetic interactions significantly prevail over chemical and thermal interactions, which typically occurs at ion energies ranging higher than several 100 eV or even higher than 1 keV [19]. However, the sputtering process efficiency in atom removal is significantly dependent upon the material sputtered. One simple schematic representation of the plasma-sputtering chamber is presented in Figure 2a.

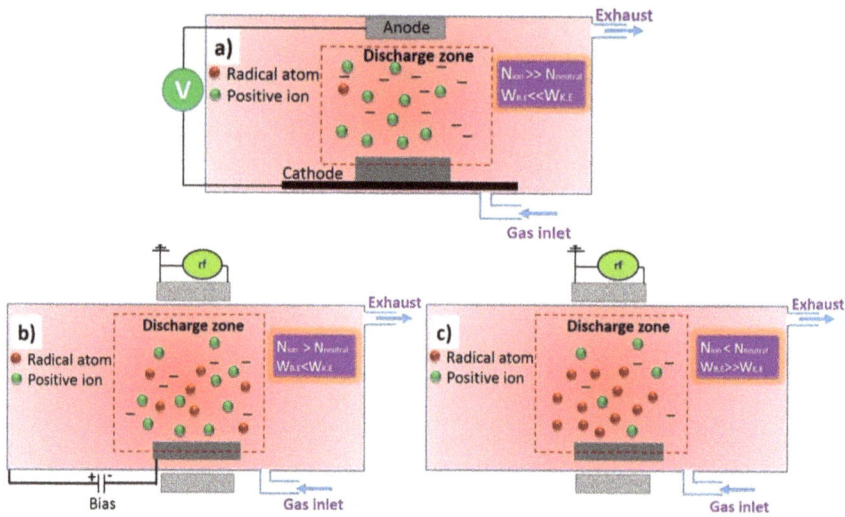

Figure 2. Schematic representation of various plasma processing systems for (**a**) sputtering; (**b**) reactive ion etching; and (**c**) highly dissociated weakly ionized plasma for chemical etching.

The major drawback of the sputtering process is its inefficiency for the surface modification of delicate materials including polymers. The problem is that the high-energy particle–surface interactions are uncontrollable and completely damage the surface mostly through cascade collisions inside the material, which provide almost no selectivity. Moving away from high-energy ions, which cause only the pure kinetic-physical interactions of ions with surfaces (via two body collisions), there is a gray zone of so-called chemical sputtering, where some chemical reactions between the incident beam and substrate occur. This process involves mostly lower energy ions (typically with energies around 100 eV) and is commonly applied to remove the surface contaminates such as hydrocarbons [20].

2.2. Reactive Ion Etching

Reactive ion etching (RIE) was mostly developed and improved by the microelectronic industry in the last few decades for trenching and patterning the surfaces. The RIE process is very diverse and uses various plasma combinations, with species ranging from merely low-energy ions (10 eV to several 100 eV) to combinations of ions with radicals, excited atoms, and electrons. These combinations are mostly generated with glow discharges, where the ratio of ($W_{K.E.}$) to ($W_{B.E.}$) is much lower than in the case of sputtering (Figure 2b). In these processes, the etching material is typically DC-biased in order to increase the effect of ion collection and acceleration towards the surface. The process is most commonly done inside a chamber with low pressure of a selected gas, where plasma is generated in a high-frequency discharge [21]. In such systems, where the surface potential is controlled, the interacting ions are directed, their energy distribution controlled by bias voltage. Nevertheless, other reactive species including neutrals inside the plasma play a significant role in determining the etching rate and etching anisotropy. Like in the case of any plasma process, the etching rate inside a reactive ion chamber is partially dependent on the discharge parameters that influence plasma properties. Therefore, the increase in the discharge power improves the etching rate, since the dissociation efficiency increases with the applied power. On the other hand, the generation of self-bias potential at reduced pressure improves the etching efficiency by energizing the particles [22].

The most reported discharge systems used for RIE are capacitively coupled RF plasma (CCP) and inductively coupled RF plasma (ICP). In CCP plasma, the ion energy varies as a function of applied power, whereas an increase in gas pressure reduces the acceleration of ions by collisions with the background gas. The ion energy flux to the surface is increased by bias voltage, which imparts a high kinetic energy to the particles. This makes complications by reducing the control or selectivity over the etching process. However, lowering the pressure also results in an increase in the collision mean free path, which restricts the existence of plasma at the point in which the mean free path approaches a value that is of the same order than the distance between the electrodes [22,23]. On the other hand, ICP is capable of generating high particle densities without any additional biasing, even at higher gas pressures [24].

The etching rate in RIE processing varies from a few nm to a few hundred nm per minute depending on the conditions used for the processing [25]. Optimization of the process is extremely important because of the problems encountered during RIE such as bowling, undercutting, or mask scattering, all of which adversely affect the aspect ratio of the processing, especially during the fabrication of circuit boards [26]. To improve the etching rate and aspect ratio, RIE is sometimes combined with other processing techniques including magnetic-field-induced beam control, lithography, ultraviolet (UV) irradiation, *etc.* [27,28].

2.3. Neutral Radical Plasma Chemical Etching

Very frequently, processes like ion sputtering and even RIE lead to thermal damages of the surface or even of bulk properties, which make them improper for the etching of temperature-sensitive materials such as polymers. In order to avoid overheating by ion bombardment or even by increased neutral gas temperature, it is wise to use mostly neutral plasma species like neutral atoms. This is achieved with so-called cold plasmas mostly generated at lower pressures (around 100 Pa) with high-frequency discharges. The high frequency applied accelerates the electrons, whereas the heavier ions cannot follow frequency, which results in a high dissociation and low ionization of the feeding gas (Figure 2c). When treatments are performed in the post-glow or after-glow regions, the discharge features practically no ions. Microwave discharges are popular for the plasma treatment of materials because they are characterized by low plasma potential, which has the advantage of insignificant ionic effects, but the neutral gas temperature is normally high. On the other hand, RF discharges have the electron temperature of about 5 eV for plasma gases such as O_2, where the observed neutral gas temperature is lower compared to microwave discharge. In such high frequency discharges, the dissociation is almost linearly enlarged with the increase in discharge power, independent of the gas flow [29].

Highly dissociated plasma is not only applicable for polymer etching, but also for other applications such as chemical reduction, nanostructuring, and plasma cleaning [30–32]. The reactions of polymers and other materials will be more elucidated in the following chapters, where different neutral atoms contribute a major part to etching through adsorption and recombination processes [33–35]. The recombination of atoms at preferential spots on the surface originates from the roughness of the surface and plays a crucial role, even in deciding the nanostructure growth [24,36]. The surface interactions and bond-breaking mechanisms during the interactions of non-equilibrium plasmas with various carbonaceous materials are also reported [37,38].

3. Applications of Plasma Functionalization and Etching of Polymers

Polymer materials cover a large segment of our material requirements for packaging, microelectronics, photonic devices, medical implants, sealing applications, water repellent coatings, thermal and electrical insulators, sensing materials, *etc.* [39–42]. Plasma-assisted surface functionalization and etching were proven to provide the desired surface energy and morphological changes to the polymeric surface [43,44]. Although various chemical treatments such as soaking in acidic medium are efficient and inexpensive methods for improving the surface energy of polymeric materials, they are associated with persisting residuals after treatment and, to a large extent, connected to environmental pollution [45,46]. Plasma modification consists of surface functionalization, which is considered to be a primary step in which surface chemistry is altered by bond breaking and the incorporation of functional groups. Thereafter, the removal of material takes place through the bonding of surface atoms with impinging radicals, which recombine and leave the surface. Through this mechanism, the organic surface contaminants or weakly bonded surface layer are removed. An alternative to plasma pre-treatment is UV irradiation, which mostly results in surface functionalization, but it is not sufficient for material etching. Moreover, wet chemical treatments as the second alternative are limited by the type of incorporated functional groups [47]. By taking these factors into consideration, plasma treatments are preferably used in polymer treatments for a wide variety of applications.

Among various applications of plasma surface treatments, improving the adhesion of various metals on polymer substrates is an important task. The major challenge in metallizing polymeric surfaces is their lower surface energy, which confronts the adhesion of metallic particles. By incorporating suitable polar functional groups by plasma exposure, this deficiency is easily solved. For instance, Pascu *et al.* presented both microwave and RF plasma treatments of polyvinylidene fluoride with N_2 and NH_3 as working gases for the incorporation of polar nitrogen containing groups including amines, nitriles, or imines. These incorporated polar functionalities then aided to increase the amount and adhesive strength of the deposited copper [48]. Metallized polymers including polyethylene terephthalate (PET) are important in the fabrication of microelectronic and photonic devices, where the intermixing of the metal and polymer showed the significant improvement of their properties when material was exposed to post-deposition annealing in Ar or Ar/O_2 plasma [49]. The treatment of the samples was done with Ar/O_2 plasma generated at low pressures (50–100 Pa) within RF discharge at power 35 W. The process increased the surface roughness and porosity, allowing the diffusion of coated Au/Ag metal from the surface into the bulk. To obtain the best adhesion, the plasma treatment conditions were optimized to minimize fracture at the interface between the metal film and polymer substrate [50,51]. The adhesion of Cu layer on polyamide also increased with O_2 plasma treatment at 200 W and 8 Pa for 3 min prior to Cu metallization. The peel strength of the deposited Cu on polyamide increased to a value of 250 g/mm, which was almost 280 times higher than that of the non-treated sample [50]. Irrespective of the gas or discharge condition used for the treatment, the metal coatings of polymers should be done immediately after the plasma surface modification in order to eradicate the drawbacks arising from the aging effects [52,53].

In addition to the improved adhesion of inorganic particles, plasma treatments have a significant role in fabricating surfaces, which are customized for the adhesion of various organic or biomolecules. Surface treatment with oxidizing plasma source gases such as O_2, NH_3, N_2, Air, or even noble gas mixtures is found efficient to enhance the surface energy and thereby adhesion of various bio materials onto the surface [54,55]. The surface roughness and polarity provided by plasma exposure are utilized to increase the adsorption of anti-blood clotting agents such as heparin onto various polymeric substrates. This operation is extended in application for manufacturing artificial organs [56]. In addition to this, the cell growth on various polymeric surfaces efficiently increases after suitable plasma pre-treatments as a simultaneous effect of improved surface contact area and the electron rich behavior. The incorporated functional groups allow for the nucleation and adsorption of desired biomaterials on the surface of the implant used by which the spatial alignment of the bio-molecules on the surface is regulated [57,58]. Another supportive application of plasma etching in the bio-medical field is plasma sterilization of surgical devices, drug packaging, or processing of other medical objects [59,60]. For example, on exposure to reactive O_2 plasma neutral radical in the afterglow, the blood proteins and microorganisms showed exponential etching rate, whereas the PET substrate exhibited only a linear etching rate [61]. These distinct etching rates help to remove toxins or other organic contaminants from the surface without affecting the surgical equipment or medical material. Further applications of plasma degradation of materials extend into novel applications such as decontamination of toxic warfare agents, the cleaning of dental devices, and even the precise removal of cancerous cells [62–64].

Besides the improved surface energy of polymeric substrates, plasma treatment is also able to incorporate low-energy functionalities on the polymeric surface. This is largely applied in the fabrication of water-repellent surfaces, which are significantly important in terms of generating better anti-aging or anti-corrosive properties of membranes for oil-water separation, and of fluid transportation control [65,66]. The self-cleaning ability of materials is expressed in terms of the water contact angle, where the surface with a contact angle above 150° is termed as the superhydrophobic surface [67]. Plasma etching efficiently creates superhydrophobic surfaces by two major mechanisms: firstly by providing sufficient roughness to the surface by the etching process and secondly by providing a sufficient number of low-energy functional groups [68]. Typically used fluorine containing plasmas have an ability to simultaneously incorporate sufficient amounts of low-energy functional groups as well as create the desired surface roughness through etching. In the cases of fluorinated polymers including Teflon (polytetrafluoroethylene—PTFE), even fluorine-free gases are sufficient enough to generate hydrophobicity merely by increasing the surface roughness by polymer etching [66,69]. An example of this are nanocone structures on Teflon surface achieved after etching the surface with O_2 plasma (25 Pa and 50 W) for 10 min. In this case, the water contact angle achieved was 134°. Such nano-featured surface is achieved also by etching anisotropy, which is created with deposits like polymer beads. The surface decorated with polystyrene (PS) beads with a 10-µm diameter prior to the plasma process can create nanocones during etching as well (Figure 3) [69]. The sufficient roughness is boosted by the additional deposition of Au nanoparticles on top of the nanocones, which leads to a superhydrophobic surface response. In certain cases, the roughness on the etched surface is influenced by the sputtered deposits from the plasma chamber wall, which act as a mask to protect the polymer but provide similar anisotropy for etching [70].

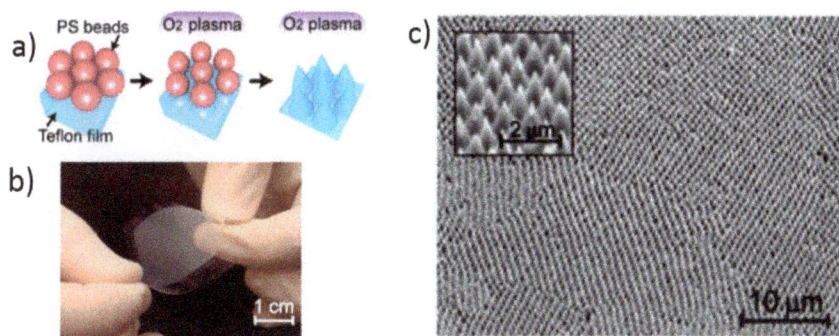

Figure 3. (a) Scheme of the fabrication process of Teflon nanocone arrays; (b) Photograph showing a macroscopic view of flexible Teflon nanocone array; (c) Scanning electron microscopy (SEM) images of the tilted nanocone array. Inset: detailed view of Teflon nanocones (Reproduced with permission from [69]. Copyright American Chemical Society, 2014).

4. The Origin of Plasma Etching Selectivity

The importance of the dry etching of polymeric materials for numerous applications such as masking or nanostructuring has increased the demand for systematic studies on the stability of various polymeric materials towards reactive plasma particles. The stability of the polymer definitely depends on the type of polymer/monomer units present (homo or copolymer, aliphatic or aromatic, crystalline or amorphous, *etc.*) as well as the energy and type of plasma particle interacting with it. Herein, the important factors are directly related to the stability of polymers towards plasma particles, which result in selective etching that will be presented and discussed.

One of the primary structural features that are considered is the aliphatic and aromatic moiety on the polymer backbone. The etching rates for aromatic polymers are relatively lower compared to aliphatic ones due to the extra energy stabilization (~36 kcal/mole), provided by the aromaticity of the rings in the polymeric chain. One more explanation to support the distinct etching rate is that the aliphatic rings degrade easily to form volatile molecules, whereas the aromatics generally form more non-volatile fragments and are not easily removed from the surface [71]. However, plasma-initiated degradation can occur on the bond–ring junctions, which are connected through relatively unstable secondary or tertiary carbon atoms. This argument can question the extra stability of the aromatic polymer chains exposed to the plasma. An appropriate argument was given by Taylor and Wolf, who stated that the extra stability of the aromatic polymers, irrespective of the position of the aromatic ring, originated from the ability of aromatic rings to quench the reactive plasma particles [72]. The quenching is typically directed through the abstraction of the hydrogen from the aromatic ring by the incoming reactive atom and the formation of a functionalized ring (Figure 4). This reaction prevents the formation of any active sites or dangling bonds on the polymer chain and prevents the chain cleavage. Additionally, the energy released during this process will be readily reduced by contributing to the vibrational energy of the polymer backbone [73]. Some early reports on the gas phase reactions of aromatic rings with oxygen plasma radicals significantly support this argument [74,75]. During such reactions, the hydroxylation was mostly directed towards the ortho position regardless of the size of the substituents on the ring due to the higher degree of neucleophilicity of oxygen radicals. This observation points out that the reactions are mostly guided through a kinetic route to give rise to reaction products that are thermodynamically less stable [74].

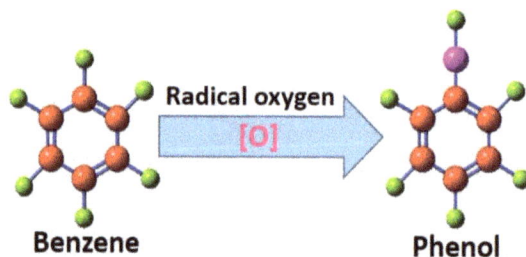

Figure 4. The schematic of reaction involved in the radical quenching by the aromatic ring to form functional group instead of ring cleavage.

Liming *et al.* provided an excellent example to present the etching efficiency and selectivity of plasma towards various chemical bonds [76]. The stability of various bonds was accurately distinguished by high-energy hydrogen radicals and ions, used to selectively etch away the edge atoms of graphene sheets and to convert them into graphene nanoribbons (GNR). As observed from atomic force microscopy (AFM) imaging, the observed etching rate for single layer graphene was 0.27 nm/min, whereas multi-layer graphene disclosed an etching rate of 0.1 nm/min (Figure 5). By keeping the surface temperature to an optimum value of 300 °C, the defect-free hydrogen terminated graphene nanoribbons were synthesized. The selectivity originated from the lower chemical stability of the edge carbon towards plasma particles compared to the atoms embedded in the mid-region of the graphene sheet [77]. In the same manner, while considering the polymeric substrates, regardless of the etching mechanism, the etching rate was mostly connected with the energy required for breaking the polymer backbone, as demonstrated by Taylor and Wolf. They compared the etching rate of different polymers (k_{rel}) with the number of chain scissions per 100 eV of energy absorbed (G_s) (Table 2). The etching rate displayed a linear relationship with the bond-breaking energy of the polymer backbone [72].

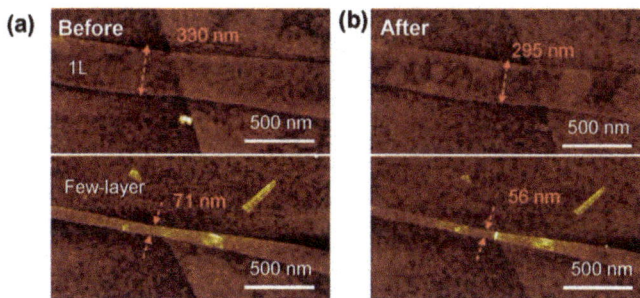

Figure 5. Atomic force microscopy (AFM) images of two small pieces of graphene (top: a monolayer (1 L) graphene strip; bottom: a few-layer graphene strip). (**a**) Before and (**b**) after selective hydrogen plasma edge etching for 60 min at 300 °C (Reproduced with permission from [76]. Copyright American Chemical Society, 2010).

Table 2. Relative rates of O_2 plasma removal (k_{rel}) and G_s-values for selected polymers (Reproduced with permission from [72]. Copyright John Wiley and Sons, 2004).

No.	Polymer	k_{rel}	G_s
1	Poly(α-methylstyrene)	1.11	0.3
2	Polyphenyl methacrylate	1.33	—
3	Polyvinyl methyl ketone	1.48	—
4	Polymethyl methacrylate (PMMA)	2.37	1.2
5	Polymethyl methacrylate-co-methacrylonitrile (94:6 mol %)	2.70	2.03
6	Polyisobutylene	3.56	4
7	Polybutene-1 sulfone	7.11	8

Another important parameter which determines the etching selectivity is the energy required for the initiation of chemical reaction (activation energy). For example, the chemical oxidation of Mg in air is highly exothermic. However, the reaction will not commence unless sufficient heat is provided such that the reactants can reach the intermediate transition state (activated complex). Once the reaction starts, the exothermic energy released will be used for the reaction to proceed. The comparison of activation energies required for the reaction between n-butane with various neutral species in the vapor state discloses that O (^1D) and F atoms do not need any activation energy for the chemical reaction to start, which reduces the selectivity towards various hydrocarbon bonds. Although O (^3P) species are reactive, the energy of activation is found to be 1 eV to react with the hydrocarbon, which indicates that this species can distinguish various chemical bonds much better than O (^1D) or F species [73,78]. Likewise, the Cl atomic species are highly reactive without any bond selectivity towards the reacting polymer due to no activation energy barrier. On the other hand, Br atoms are relatively non-reactive due to an energy barrier of around 1.8 eV. Moreover, the etching rate can also be controlled through the combination of various reactive species, which can be regulated by controlling the feeding gas composition and discharge parameters. In one recent report, it was demonstrated that adding O_2 into CF_4 plasma increased the etching rate [79]. This phenomenon was due to the involvement of excited oxygen radicals in the electron impact dissociation of CF_4, which leads to a higher dissociation and more F atomic species [80].

Hegemann *et al.* compared the etching rates for various polymers with different chemical structures inside the noble gas-generated plasmas and concluded that the polymer backbones that contained oxygen functionalities etched away relatively faster. For example, polymethyl methacrylate (PMMA), PET, *etc.* showed relatively higher etch rates compared to the hydrocarbon polymers inside Ar plasma generated at 300 W and a pressure of 20 Pa (Figure 6) [81]. The explanation for the higher etch rates for functionalized polymers was due to the secondary reactive oxygen atomic species released into the system as a result of polymer bond cleavage. More surprisingly, polypropylene (PP) exhibited a lower etching rate compared to PS, which was in contradiction to the hypothesis of radical quenching mechanism proposed for the extra stability of aromatic polymers. However, this was explained by the efficient cross-linking on the PP surface upon exposure to plasma due to the side chain activation [81].

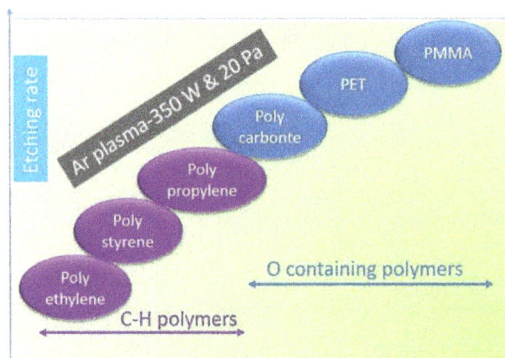

Figure 6. Etching rates for various polymer substrates in Ar plasma based on [81].

Plasma surface interactions have a strong correlation with the crystallinity of the polymer used, where the etching rate is typically reduced with the increase in crystallinity [82]. The reason for this can be different neutral atom recombination probabilities on surfaces. For example, the neutral O atom recombination probability is 8.3×10^{-4}–2.1×10^{-4} for amorphous PET or 4×10^{-4}–2.5×10^{-5} for semi-crystalline PET [83]. Thus, the extension of interactions between the polymer and the neutral atomic species that induce etching differ with respect to crystallinity, which cannot hold for charged species. The ion flux to the surface is independent of the crystallinity. The increased value of the neutral atom recombination coefficient provided higher contribution to the surface temperature which induced more deformation to the amorphous material, as reflected by the roughness measurements before and after plasma exposure [83–85]. The treatment of LDPE (low density polyethylene) and HDPE (high density polyethylene) inside CF_4 plasma revealed similar results. The etching rate for LDPE was 1.5 nm/min more than that of the HDPE inside CF_4 plasma generated by capacitively coupled RF discharge, where the morphological changes also displayed significant differences with respect to crystallinity [86]. Nair *et al.* recently reported the removal of an amorphous carbon layer from the multiwalled carbon nanotube (MWCNT) with low-energy ions, radicals and metastable species, where the crystalline phase remained relatively intact. The extra stability provided by the crystallization energy additionally supported the etch-resistant behavior of crystalline polymers [87]. From all of the above-mentioned results, the relative etching rates for various polymeric substrates inside the widely used oxygen plasma are summarized in Figure 7.

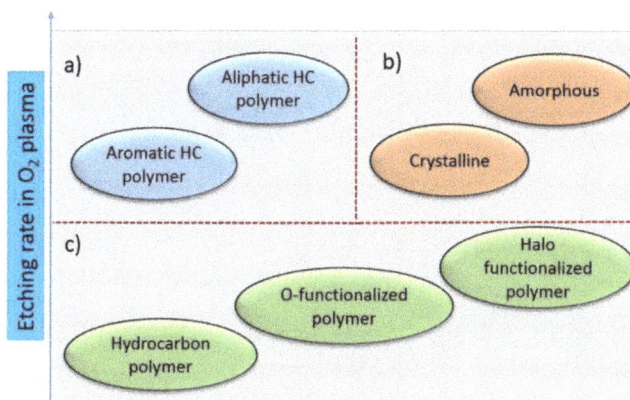

Figure 7. General schematic of the etching rate for various types of polymeric materials in O_2 plasma. (a) Dependence of etching rate on the aliphatic/aromatic behavior of the monomer units; (b) Etching rate dependence on the crystallinity; (c) Functionality dependence of the polymer with etching rate.

An important class of polymers that needs to be separately addressed is that of block copolymers, since they are largely used in block copolymer lithography for the semiconductor industry. The chemical dissimilarities between the two polymer domains permit the selective etching of one component over the other, when a suitable treatment method is used. After the selective etching of the less stable polymer, the unaffected component will form the template pattern. One of the most widely used copolymers is the PS/PMMA system, where both of the components have distinct chemical stabilities. For achieving spherical and cylindrical structures, PMMA is easily removed with a suitable wet chemical treatment [88]. However, the wet chemical etching is limited due to capillary forces when the polymers possess lamellar arrangement. In order to avoid these problems, plasma selective etching is implemented and used for a number of copolymer systems: for example, polystyrene–poly ferrocenylisopropylmethylsilane diblock copolymer, styrene–butadiene–styrene (SBS) triblock copolymer, polystyrene–polydimethylsiloxane block copolymer, polyhydroxybutyrate-co-hydroxyvalerate, *etc.* [89–91]. The most fundamental object while choosing the copolymer for plasma etch patterning is the selection of proper blocks that have distinct etching rates towards various plasma particles [92].

The oxidation probability of the polymer is independent of the density of plasma particle species in the proximity of the samples, but the etching still depends on plasma particle properties and their fluxes to surface [93]. In many cases, the etch-resistant properties of polymer materials are dependent on the type of discharge used, where the difference in the etching rate originates from the energies and flux of various reactive species including neutral atoms and ions generated [37,94,95]. The etching rate is also well correlated to the surface temperature originated from the ion bombardment, neutral atom recombination, and exothermic carbon oxidation [96–98]. In order to achieve the highest etching rates, a higher surface temperature is always preferred. However, the surface temperatures above the glass transition temperature of the polymer can critically affect the bulk properties. Due to this, the radical and ion flux should be optimized, whereas in many cases the pulse mode plasma treatment over continuous mode will be more appropriate to control the temperature related effects [99–101].

5. Applications of Plasma Selective Etching of Composite Materials

Plasma selective etching is utilized in many applications for material fabrication especially in micro- and nano-structuring. One of the important examples is the transplantation of nanostructures like nanotubes. Li *et al.* demonstrated a method for growing CNTs inside Si trench with Ni catalysts.

The challenge faced was the transfer of the designed nanotubes onto suitable receptors. This difficulty was solved by extracting the nanotubes with an epoxy matrix in the form of pellets, followed by oxygen plasma selective etching of the epoxy matrix in order to release well aligned and densely packed CNT arrays [102]. By taking advantage of the difference in etching selectivity of various chemical bonds towards plasma, the metal single-walled carbon nanotube (m-SWCNT) were also obtained. One of the challenges for creating m-SWCNTs is the higher chemical etching rate of m-SWCNTs compared to semi-conducting single-walled carbon nanotubes (s-SWCNT) [103]. However, by taking advantage of the weaker stability of smaller diameter s-SWCNT due to the C–C bond bending, a mild hydrogen plasma treatment yielded m-SWCNT from a mixture of m- and s-SWCNT [104]. This etching selectivity is only diameter-dependent, unlike the m- or s- character of SWCNT, and could yield a 100% recovery and easy scaling up of one type of SWCNT from the mixture [105].

Graphene-based electrodes are used in wide variety of semiconducting devices such as field effect transistors, sensors, supercapacitors, *etc.* [106]. In the transistor applications, one of the most important parameters is the gap dimension between the electrodes. Graphene sheets, which are separated by micro- or nanoscale distance, played a significant role in the fabrication of numerous electronic devices. The fabrication of such devices is extremely hard by wet chemical treatment due to the lack of precision in etching over nanoscale dimensions. As demonstrated by Liao *et al.*, plasma-assisted mask etching can provide precise control over the gap between graphene sheets ranging from a few micrometers to hundreds of nanometers [107]. Furthermore, the electrical properties of graphene sheets are highly dependent on their specific surface area. For this, graphene sheets can be converted into different porous forms including graphene nanomesh, crumpled graphene, folded graphene, and graphene foam in order to exploit the exceptionally large surface area [108–110]. Such high surface area graphene sheets are also easy to obtain by means of polymeric or metallic mask-assisted plasma selective etching. In a presented example, the spin coating of PMMA on graphene-supported silica, when covered with porous anodized aluminum oxide resulted in a suitably masked surface for plasma etching. Treatment of this composite with O_2 plasma for 30 s resulted in pores with diameters of 67 nm on the graphene surface, which produces the graphene mesh with a very high surface area of 100 μm^2 [111]. The origin of etching selectivity during the processing was based on the tolerance of the Al mesh towards plasma compared to the PMMA layer. The observed difference in the etching rates of PS and P(S-r-MMA-r-GMA) (a random copolymers of styrene, methyl methacrylate, and glycidyl methacrylate) was 1.17 and 1.42 nm/s, respectively, inside O_2 plasma (50 W, 10 sccm, 1.3 Pa), compared to 0.76 and 0.96 nm/s, respectively, inside the CHF_3/O_2 gas mixture (300 W, 45 sccm CHF_3 and O_2 5 sccm, 8 Pa). Exploiting this, the patterned copolymer template was fabricated, which provided an easy route for creating nanoperforations on graphene sheets. Such prepared graphene with a large surface area has been used in electronic and sensing devices [112]. Additionally, the combination of holography and photoresist-assisted O_2 plasma selective etching was demonstrated for providing a low-cost production route to synthesize graphene nanomesh [113]. Additional examples are also seen where self-assembled masks are used for the perforation of graphene sheets during selective etching [114]. Such masks provide controlled exposure of the graphene layer to the plasma reactive species and yield precise mesh sizes and ribbon widths between successive meshes. This allows control over the electronic characteristics of graphene [115]. During the etching of such graphene composite surfaces, the selective edge oxidation and O_2 physisorption on the inner walls of the mesh add secondary benefits to chemo-sensing applications [116].

Another particular application of selective plasma modification and etching was to improve gas permeation properties of composite membranes [117]. After plasma treatment with corona discharge, the polyamide 6/polyethersulfone (PA6/PES) composite membrane exhibited significant improvement in separation of CO_2, N_2, and O_2 by incorporating functional groups and increasing the surface roughness [118]. Unfortunately, the changes induced to the distinct composite components were not studied individually. When a polybutadiene/polycarbonate (PB/PC) membrane was exposed to the Cl-containing plasma, it was observed that the modified surface chemistry had very little influence, whereas the separation towards N_2 and O_2 were dominantly affected by the variation in surface morphology [119]. Other than for permeation applications, selective plasma-assisted mask O_2/Ar etching of Nafion membrane has been used in the preparation of patterned metal-polymer composite membranes for robotics and other applications [120].

Unzipping CNTs under plasma exposure is known to ease the preparation of novel materials including hollow CNTs or GNR. To make such preparative methods easier, the simplest method is to expose CNTs embedded in a polymer matrix to reactive plasmas. Exposure of the vertically aligned CNT in the polymer matrix to plasma opened the tips of the embedded CNTs. This route also included the removal of the polymer matrix at higher rates and then exposing the tube tips and further opening of the CNT tips [121]. In another relevant example, plasma selective etching enabled the successful synthesis of single-layer GNR from CNTs embedded in PMMA composite. The method provided excellent control over edge smoothness and uniformity in width for the fabricated ribbon structures [122].

Due to its simplicity, and the higher rate of dimension tolerance compared to wet chemical etching procedures, plasma etching is well exploited in the semiconducting industry for printing integrated circuit boards. One of the disadvantages often given for plasma processing is the high production cost due to expensive low-pressure plasma systems, which can be bridged only with long-term operation. Silicon wafers, as one of the most used materials in the semiconductor industry, is efficiently etched mostly by halogen-containing plasmas or by their mixture with other gases in the presence of a suitable etch mask. The etch mask on the surface generates an unbalanced etching rate of the surface and creates desired patterns on the wafer [123]. For achieving the etching selectivity between the semiconducting layer and the photoresist, the control of the plasma species inside is needed. By adding certain gases, it is possible to scavenge the radicals and ions, which presents an alternative route to controlling the species inside the plasma and the corresponding etching selectivity, without changing the discharge parameters [124]. One of the major drawbacks of RIE technology is the non-uniform etching on the sidewalls. It operates in such a way that the radicals get accumulated nearer to the edge of the wafer instead of the center, which increases the etch rate towards the edges. Additionally, the ion current is favorably drawn over the wafer edge, and the etch rates for the silicon wafer show an increase from the center to the edge and the consequent edge breakings [125]. Performance level and many of the disadvantages of RIE processing such as anti-notch performances, quality of the profile, surface finishing, *etc.* are tuned by adjusting the applied frequency or pulsing [125]. Even in the presence of these disadvantages, plasma etching is far better than other conventional methodologies.

An important application of the diverse material etching rates in plasma is used to deal with the nanostructuring of the polymeric substrates. Polymer-based 1D and 2D nanostructures have potential applications in sensing and energy devices [126,127]. Although template-assisted synthetic methods are well-known, controlling the dimensions during template removal still represents an obstacle. Plasma-assisted mask etching has proven to be an alternative to overcome these problems. An example is the exposure of metallic nanoparticles deposited on polymer surfaces to reactive gas plasma, which yields the dense polymeric NWs. The metallic coating or, say, metallic material islands (nanodots) merely acted as an etch mask to provide a rough surface to mobilize the NW growth [128]. For this purpose, a mixture of Ar, O_2, and CF_4 gases was leaked into the inductively coupled plasma chamber at constant flow ratios of 15, 10, and 30 sccm. The plasma discharge was generated at 400 W where an additional 100 W was used for biasing to accelerate and direct the ions to the surface. In this way, the polymeric NWs with an aspect ratio of up to 700 were produced on the surface of a large number of polymers including PET, Kapton, Dura film, PS, and polydimethyl siloxane (PDMS). The length of the NWs exhibits a linear trend with the exposure time, whereas the diameter remains constant at a value around 100 nm for a given thickness of the mask on the surface. On the other hand, the density of the NWs is strongly influenced by the thickness of the coated metal. The scanning electron microscopy (SEM) images of NW with respect to various thicknesses of the mask and the relationship of the etching rate with the plasma exposure time are presented in Figure 8.

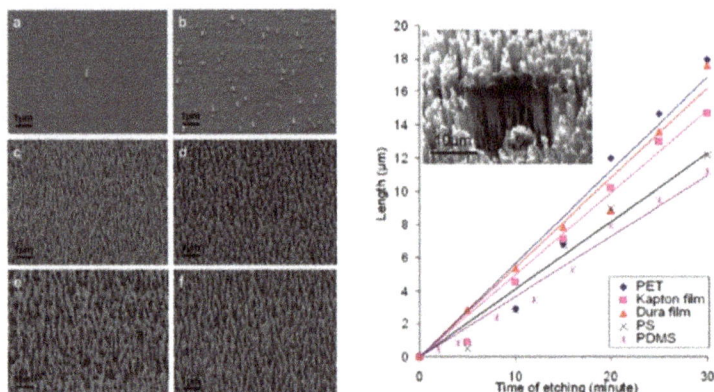

Figure 8. SEM images of the density-controlled fabrication of polymer nanowire (NW) arrays of Kapton by covering the initial surface with (**a**) 0.75; (**b**) 1.5; (**c**) 3; (**d**) 4.5; (**e**) 10; and (**f**) 15 nm of Au before inductively coupled plasma (ICP) etching. The graph represents the length-controlled growth of NWs of polyethylene terephthalate (PET), Kapton film, Durafilm, polystyrene (PS), and polydimethyl siloxane (PDMS). The inset is a SEM image of a NW array on Durafilm after 30 min of etching (Reproduced with permission from [128]. Copyright American Chemical Society, 2009).

In many of the presented examples, the surface nanostructuring is achieved by using an appropriate mask/template. Nevertheless, it was not necessary to introduce a mask prior to processing. A typical example is found where simultaneous plasma-enhanced reactive ion synthesis and etching (SPERISE) is employed. This technique was first used during the treatment of a Si wafer inside HBr-O_2 gas plasma (Figure 9a). In the primary step, the halogen species interact with the surface, and Si atoms are released into the vapor phase. This excited Si is then combined with the O and Br reactive species to form an etch-resistant silicon oxy bromide complex, which is deposited back on the surface. These deposits then act as masks and support the non-uniform etching on the surface. As a result of this processing, the mushroom-like Si NWs are formed, where an etch mask is left on the top (Figure 9b). The removal of the etch mask is done later with relatively simple mild chemical treatment [129].

Another application of plasma selective etching includes the study of filler dispersion inside the composite matrix. The filler dispersion inside the polymer matrix is the deciding factor for its bulk properties including ductility, hardness, impact resistance, *etc.* Conventional methods such as cryogenic breaking are generally used to monitor the dispersion of the fillers [130,131]. However, this method lacks precision for the determination of dispersion due to the possible displacement of the fillers during the distortion of the matrix. To overcome this, dissociated plasma etching is applied where the surface polymer layer is selectively removed to expose the embedded fillers. A fast and selective removal of the surface polymer from the composite materials including powder coatings and paint films with the dissociated oxygen plasma followed by SEM imaging enables the monitoring of the filler dispersion [132–135]. The same strategy allowed for the optimization of the bonding process efficiency of powder coatings with other materials, where bonding is enabled through disclosed but still embedded fillers. Due to the inhomogeneity in the plasma etching of various components inside the composite material, the resulting surface acquires relatively high surface roughness due to the exposed fillers. Along with suitable plasma functionalities, increased surface roughness is gained, which additionally improves the metallization of the surface [136,137]. Similarly, the fabrication of counter electrodes for solar cell application was done from a PP/CNT nanocomposite, where O_2 plasma reactive ion etching was used to remove the thick protruding layer of PP on the surface to expose the embedded CNTs. As a result, the charge transfer resistance of the composite surface is diminished to much lower values [138].

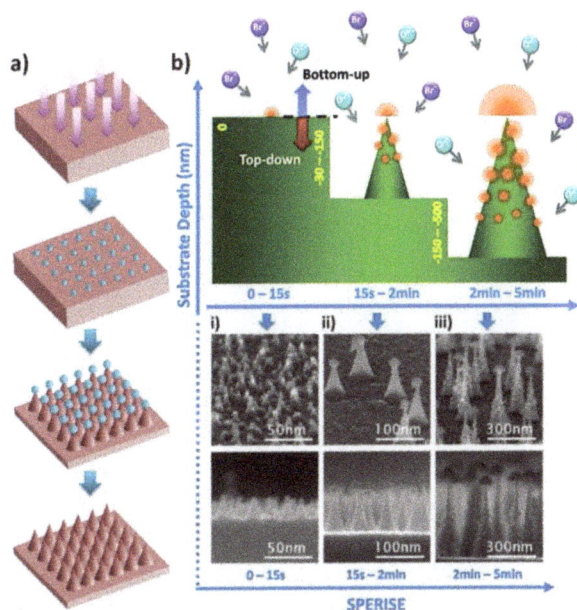

Figure 9. Simultaneous plasma enhanced reactive ion synthesis and etching (SPERISE) process and Si nanocone formation mechanism. (**a**) Process flow of the nanomanufacturing process: Pseudo randomly distributed silicon oxybromide nanodots are synthesized on the planar silicon substrate surface in the first few seconds of the SPERISE process. The oxide nanodots grow to hemispheres by a phase-transition nucleation process and act as a protective nanomask for the simultaneous reactive ion etching of the silicon underneath. Depending on the growth rate of the oxide hemispheres and the crystalline structures of the silicon substrates, nanocones with different aspect ratios are formed. The silicon oxybromide nanohemispheres on top of the nanocones are removed by wet etching; (**b**) Detailed schematic drawing of the three typical stages in the SPERISE process: Bromine and oxygen reactive ions interact with silicon to form synthesized oxide hemisphere and dots (orange) and etched silicon cone structure (green). Both the illustrations and corresponding SEM images at (i) 0–15 s; (ii) 15 s–2 min; and (iii) 2–5 min in the SPERISE process manifest this unique nanomanufacturing method (Reproduced with permission from [129]. Copyright American Chemical Society, 2011).

The applications of plasma selective etching are further extended for achieving improved bio- and chemo-sensing properties of semiconducting materials. As shown in one recent report, the gas-sensing properties of a composite membrane prepared by electrospinning process from the aq. solution of polyvilnyl alcohol (PVA) and SnO_2 was improved by O_2 plasma treatment. The resulting SnO_2 material showed a notable sensing response towards very low concentrations of ethanol vapor (~1 ppb). The enhanced sensitivity was featured because of the high specific surface area of the ripple-like structures obtained after plasma treatment followed by annealing at 500 °C [139]. However, the metal oxide nanodevices are less preferred due to their inability to operate at room temperature for sensing applications that could be replaced by conducting polymers, carbon allotropes or their suitable composites [140–142]. The sensing ability of such materials, especially in the context polymeric composites are further improved by plasma modification of the surface. As demonstrated by Raghu and co-workers, plasma modifications of the MWCNT composite with conducting polymers indicated greater effects on sensitivity and selectivity towards various volatile organic compounds [143]. However, a better understanding of how plasma improves the sensitivity and selectivity is still to be studied extensively for improving the sensor properties. While in the case of bio-sensors, the sensitivity is found to be controlled by the covalent attachment of the bio-molecule to the sensing material. For this application, the removal of the polymer matrix and the functionalization of the exposed MWCNTs on the surface of a pristine PS/MWCNT composite yielded an opportunity to covalently attach antibodies and to fabricate advanced immunosensors [144].

One of the most recent applications of plasma chemical etching was the improvement of the insulation properties of polymeric composites. The insulating properties of various polymeric materials are standardized in terms of comparative tracking index (CTI) as per international electrotechnical commission (IEC) grading. The origin of poor insulating properties of composite materials is an aftereffect of the polymer charring on the surface at a high voltage electric arc. This issue is commonly tackled by adding suitable fillers, which forms little or no char on the surface [145]. However, this is inadequate due to the thin polymer layer on top of the embedded fillers. The simplest solution to overcome this situation is the selective removal of the surface polymer layer by an optimized plasma discharge. The cold O_2 plasma removal of the surface polymer from the glass embedded phenolic resin composite was demonstrated to improve CTI performances up to 56% (Figure 10) [101]. The performance level increased with the decrease in the surface polymer content along the prolonged plasma treatment times.

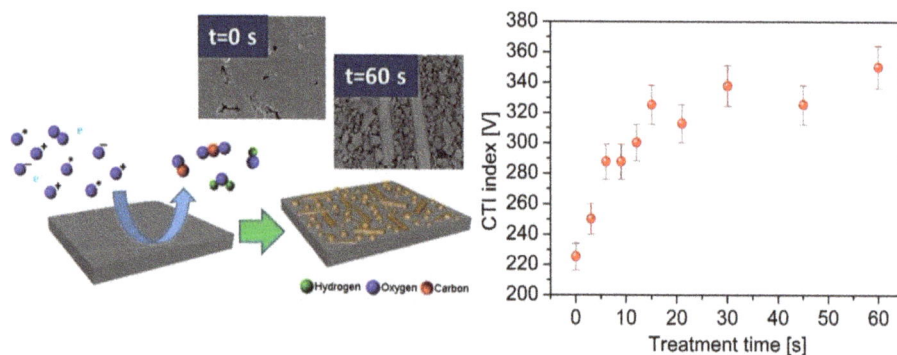

Figure 10. Plasma surface interaction of the glass-filled composite with corresponding SEM images for non-treated and plasma-treated samples for 60 s. The graph represents the variation of comparative tracking index (CTI) performance with plasma exposure time (Reproduced with permission from [101]. Copyright Royal Society of Chemistry, 2015, Year.").

6. Conclusions and Research Challenges to Tackle

This review attempted to outline the interdisciplinary applications of plasma etching and the selective etching of polymer-based materials from different branches of science. Moreover, it looks at the origin of selectivity and attempts to find answers bringing together an understanding of plasma properties with very diverse results of plasma–surface interactions through the note to nanomaterials. Among various plasma modifications, neutral dense and ion free cold plasmas are have been found to be efficient, especially for treating delicate polymeric and biomaterials to avoid unwanted surface damages and thermal effects. The plasma-induced functionalization and etching of polymer substrates is preferred to wet chemical etching and UV irradiation for designing the surface chemistry, surface morphology, and surface energy. These plasma induced modifications enable the attachment of various materials and biomolecules onto the surface. Additionally, plasma-induced hydrophobization of the surface can effectively increase the water-resistant behavior of the surface to improve anti-aging, anti-fouling, and corrosion-resistant properties. Such improvements in the surface properties are directly connected to structural and morphological changes.

The high-energy particles inside the plasma are able to distinguish various bond types under controlled process parameters. Utilizing the difference in the chemical stability of various bond types on the same molecule, delicate materials such as GNRs have been fabricated. The comparison of etching rate for different classes of polymeric materials has revealed that the difference in the etch rate is connected with both physical and chemical properties of the material. The presence of aromatic moieties has reduced the etching rate by radical quenching, whereas the surface cross-linking has reduced the etching rate for branched polymers. Generally, the etch rates are lower for hydrocarbon polymers compared to functionalized macromolecules. Additionally, a higher extension of crystallinity has reduced the plasma etch rates for various polymers, presented through reports on a few examples, namely, PET, LDPE, and HDPE. Thus, existing differences in the etching rates for various polymeric substrates have been employed in the surface structuring of block copolymers in applications, especially for the semiconductor industry. The pronounced applications of plasma selective etching of composite materials have been found to be efficient for reducing the dimensionalities of materials such as CNT or graphene sheets for advanced applications in electronics, electrical, and sensing devises. Furthermore, the preferential etching of composite surfaces has been effectively used for a simple and low-cost synthesis of polymer NWs with a controlled aspect ratio.

The plasma etching rate and etching selectivity of micro/macro molecules are well connected with the strength of the available bonds in the material. However, the side reactions such as radical quenching and surface cross-linking significantly affect the etching rate. Thus, the optimization of the process parameters is essential for achieving etching selectivity in different systems. Plasma is used as a single operating tool in only one/a few step(s) throughout multistep processes for demanding applications. More frequently used plasma has been found as one of the steps in multi-processing. This is clearly visible in many of the discussed examples, including the synthesis of GNR, graphene nanomesh, patterning the surface of block copolymers, *etc.* In some of the presented examples, the involved wet chemistry had slight adverse effects on the properties of the final product. An additional challenge in front of the plasma community is the pinpoint control of various plasma species and their reactions on the atomic or molecular level. Such developments are important, especially for such remarkable applications as plasma nanoscience and plasma medicine.

Acknowledgments: This work was partially funded by the Slovenian Research Agency (ARRS) projects L2-6769 and program P2-0082. Harinarayanan Puliyalil would like to thank G. Filipič for valuable discussions, and the Jožef Stefan International Postgraduate School (MPŠ) for the grant from the Innovative Scheme for Research.

Conflicts of Interest: The authors declare no conflict of interest. The founding sponsors had no role in the design of the study; in the collection, analyses, or interpretation of data; in the writing of the manuscript, and in the decision to publish the results.

References

1. Ostrikov, K.; Cvelbar, U.; Murphy, A.B. Plasma nanoscience: Setting directions, tackling grand challenges. *J. Phys. D* **2011**, *44*. [CrossRef]
2. Zhang, Z.; Han, S.; Wang, C.; Li, J.; Xu, G. Single-walled carbon nanohorns for energy applications. *Nanomaterials* **2015**, *5*, 1732–1755. [CrossRef]
3. Bo, Z.; Yang, Y.; Chen, J.; Yu, K.; Yan, J.; Cen, K. Plasma-enhanced chemical vapor deposition synthesis of vertically oriented graphene nanosheets. *Nanoscale* **2013**, *5*, 5180–5204. [CrossRef] [PubMed]
4. Filipič, G.; Cvelbar, U. Copper oxide nanowires: A review of growth. *Nanotechnology* **2012**, *23*. [CrossRef] [PubMed]
5. Guo, S.; Dong, S. Graphene nanosheet: Synthesis, molecular engineering, thin film, hybrids, and energy and analytical applications. *Chem. Soc. Rev.* **2011**, *40*, 2644–2672. [CrossRef] [PubMed]
6. Ashik, U.P.M.; Wan Daud, W.M.A.; Abbas, H.F. Production of greenhouse gas free hydrogen by thermocatalytic decomposition of methane—A review. *Renew. Sustain. Energy Rev.* **2015**, *44*, 221–256. [CrossRef]

7. Mariotti, D.; Mitra, S.; Svrcek, V. Surface-engineered silicon nanocrystals. *Nanoscale* **2013**, *5*, 1385–1398. [CrossRef] [PubMed]
8. Park, S.H.; Chae, J.; Cho, M.-H.; Kim, J.H.; Yoo, K.-H.; Cho, S.W.; Kim, T.G.; Kim, J.W. High concentration of nitrogen doped into graphene using N_2 plasma with an aluminum oxide buffer layer. *J. Mater. Chem. C* **2014**, *2*, 933–939. [CrossRef]
9. Meena, J.S.; Chu, M.-C.; Chang, Y.-C.; You, H.-C.; Singh, R.; Liu, P.-T.; Shieh, H.-P.D.; Chang, F.-C.; Ko, F.-H. Effect of oxygen plasma on the surface states of ZnO films used to produce thin-film transistors on soft plastic sheets. *J. Mater. Chem. C* **2013**, *1*, 6613–6622. [CrossRef]
10. Pumera, M. Heteroatom modified graphenes: Electronic and electrochemical applications. *J. Mater. Chem. C* **2014**, *2*, 6454–6461. [CrossRef]
11. Kumar, A.; Ann Lin, P.; Xue, A.; Hao, B.; Khin Yap, Y.; Sankaran, R.M. Formation of nanodiamonds at near-ambient conditions via microplasma dissociation of ethanol vapour. *Nat. Commun.* **2013**, *4*. [CrossRef] [PubMed]
12. Attri, P.; Arora, B.; Choi, E.H. Utility of plasma: A new road from physics to chemistry. *RSC Adv.* **2013**, *3*, 12540–12567. [CrossRef]
13. Meyyappan, M. A review of plasma enhanced chemical vapour deposition of carbon nanotubes. *J. Phys. D* **2009**, *42*. [CrossRef]
14. Meyyappan, M. Plasma nanotechnology: Past, present and future. *J. Phys. D* **2011**, *44*. [CrossRef]
15. Vasilev, K.; Griesser, S.S.; Griesser, H.J. Antibacterial surfaces and coatings produced by plasma techniques. *Plasma Process. Polym.* **2011**, *8*, 1010–1023. [CrossRef]
16. Aria, A.I.; Lyon, B.J.; Gharib, M. Morphology engineering of hollow carbon nanotube pillars by oxygen plasma treatment. *Carbon* **2015**, *81*, 376–387. [CrossRef]
17. Coburn, J.W.; Winters, H.F. Ion- and electron-assisted gas-surface chemistry—An important effect in plasma etching. *J. Appl. Phys.* **1979**, *50*, 3189–3196. [CrossRef]
18. Denes, F.S.; Manolache, S. Macromolecular plasma-chemistry: An emerging field of polymer science. *Prog. Polym. Sci.* **2004**, *29*, 815–885. [CrossRef]
19. Bohlmark, J.; Lattemann, M.; Gudmundsson, J.T.; Ehiasarian, A.P.; Aranda Gonzalvo, Y.; Brenning, N.; Helmersson, U. The ion energy distributions and ion flux composition from a high power impulse magnetron sputtering discharge. *Thin Solid Films* **2006**, *515*, 1522–1526. [CrossRef]
20. Smith, T. Sputter cleaning and etching of crystal surfaces (Ti, W, Si) monitored by auger spectroscopy, ellipsometry and work function change. *Surf. Sci.* **1971**, *27*, 45–59. [CrossRef]
21. Oehrlein, G.S. Reactive-ion etching. *Phys. Today* **1986**, *39*, 26–33. [CrossRef]
22. Lee, H.; Oberman, D.B.; Harris, J.S. Reactive ion etching of GaN using CHF_3/Ar and C_2ClF_5/Ar plasmas. *Appl. Phys. Lett.* **1995**, *67*, 1754–1756. [CrossRef]
23. Levchenko, I.; Keidar, M.; Xu, S.; Kersten, H.; Ostrikov, K. Low-temperature plasmas in carbon nanostructure synthesis. *J. Vac. Sci. Technol. B* **2013**, *31*. [CrossRef]
24. Ostrikov, K.; Levchenko, I.; Cvelbar, U.; Sunkara, M.; Mozetic, M. From nucleation to nanowires: A single-step process in reactive plasmas. *Nanoscale* **2010**, *2*, 2012–2027. [CrossRef] [PubMed]
25. Tachi, S.; Tsujimoto, K.; Okudaira, S. Low-temperature reactive ion etching and microwave plasma etching of silicon. *Appl. Phys. Lett.* **1988**, *52*, 616–618. [CrossRef]
26. Coburn, J.W.; Winters, H.F. Conductance considerations in the reactive ion etching of high aspect ratio features. *Appl. Phys. Lett.* **1989**, *55*, 2730–2732. [CrossRef]
27. Cybart, S.A.; Roediger, P.; Ulin-Avila, E.; Wu, S.M.; Wong, T.J.; Dynes, R.C. Nanometer scale high-aspect-ratio trench etching at controllable angles using ballistic reactive ion etching. *J. Vac. Sci. Technol. B* **2013**, *31*. [CrossRef]
28. Zeze, D.A.; Cox, D.C.; Weiss, B.L.; Silva, S.R.P. Lithography-free high aspect ratio submicron quartz columns by reactive ion etching. *Appl. Phys. Lett.* **2004**, *84*, 1362–1364. [CrossRef]
29. Mozetič, M.; Vesel, A.; Monna, V.; Ricard, A. H density in a hydrogen plasma post-glow reactor. *Vacuum* **2003**, *71*, 201–205. [CrossRef]
30. Filipič, G.; Baranov, O.; Mozetič, M.; Ostrikov, K.; Cvelbar, U. Uniform surface growth of copper oxide nanowires in radiofrequency plasma discharge and limiting factors. *Phys. Plasmas (1994-Present)* **2014**, *21*. [CrossRef]

31. Wang, A.; Qin, M.; Guan, J.; Wang, L.; Guo, H.; Li, X.; Wang, Y.; Prins, R.; Hu, Y. The synthesis of metal phosphides: Reduction of oxide precursors in a hydrogen plasma. *Angew. Chem.* **2008**, *120*, 6141–6143. [CrossRef]

32. Vratnica, Z.; Vujosevic, D.; Cvelbar, U.; Mozetic, M. Degradation of bacteria by weakly ionized highly dissociated radio-frequency oxygen plasma. *IEEE Trans. Plasma Sci.* **2008**, *36*, 1300–1301. [CrossRef]

33. Vesel, A.; Mozetic, M. Surface functionalization of organic materials by weakly ionized highly dissociated oxygen plasma. *J. Phys. Conf. Ser.* **2009**, *162*. [CrossRef]

34. Hartney, M.A.; Greene, W.M.; Soane, D.S.; Hess, D.W. Mechanistic studies of oxygen plasma etching. *J. Vac. Sci. Technol. B* **1988**, *6*, 1892–1895. [CrossRef]

35. Pearton, S.J.; Norton, D.P. Dry etching of electronic oxides, polymers, and semiconductors. *Plasma Process. Polym.* **2005**, *2*, 16–37. [CrossRef]

36. Filipič, G.; Baranov, O.; Mozetič, M.; Cvelbar, U. Growth dynamics of copper oxide nanowires in plasma at low pressures. *J. Appl. Phys.* **2015**, *117*. [CrossRef]

37. Cvelbar, U. Interaction of non-equilibrium oxygen plasma with sintered graphite. *Appl. Surf. Sci.* **2013**, *269*, 33–36. [CrossRef]

38. Cvelbar, U.; Mozetič, M.; Junkar, I.; Vesel, A.; Kovač, J.; Drenik, A.; Vrlinič, T.; Hauptman, N.; Klanjšek-Gunde, M.; Markoli, B.; *et al.* Oxygen plasma functionalization of poly(*p*-phenilene sulphide). *Appl. Surf. Sci.* **2007**, *253*, 8669–8673. [CrossRef]

39. Wang, K.; Zhang, X.; Zhang, X.; Yang, B.; Li, Z.; Zhang, Q.; Huang, Z.; Wei, Y. Fabrication of cross-linked fluorescent polymer nanoparticles and their cell imaging applications. *J. Mater. Chem. C* **2015**, *3*, 1854–1860. [CrossRef]

40. Tsougeni, K.; Papageorgiou, D.; Tserepi, A.; Gogolides, E. "Smart" polymeric microfluidics fabricated by plasma processing: Controlled wetting, capillary filling and hydrophobic valving. *Lab Chip* **2010**, *10*, 462–469. [CrossRef] [PubMed]

41. Yang, X.; Zhou, G.; Wong, W.-Y. Recent design tactics for high performance white polymer light-emitting diodes. *J. Mater. Chem. C* **2014**, *2*, 1760–1778. [CrossRef]

42. Uysal Unalan, I.; Cerri, G.; Marcuzzo, E.; Cozzolino, C.A.; Farris, S. Nanocomposite films and coatings using inorganic nanobuilding blocks (NBB): Current applications and future opportunities in the food packaging sector. *RSC Adv.* **2014**, *4*, 29393–29428. [CrossRef]

43. Vlachopoulou, M.-E.; Kokkoris, G.; Cardinaud, C.; Gogolides, E.; Tserepi, A. Plasma etching of poly(dimethylsiloxane): Roughness formation, mechanism, control, and application in the fabrication of microfluidic structures. *Plasma Process. Polym.* **2013**, *10*, 29–40. [CrossRef]

44. Jacobs, T.; De Geyter, N.; Morent, R.; Desmet, T.; Dubruel, P.; Leys, C. Plasma treatment of polycaprolactone at medium pressure. *Surf. Coat. Technol.* **2011**, *205*, S543–S547. [CrossRef]

45. Charbonnier, M.; Romand, M. Polymer pretreatments for enhanced adhesion of metals deposited by the electroless process. *Int. J. Adhes. Adhes.* **2003**, *23*, 277–285. [CrossRef]

46. Yu, H.-D.; Regulacio, M.D.; Ye, E.; Han, M.-Y. Chemical routes to top-down nanofabrication. *Chem. Soc. Rev.* **2013**, *42*, 6006–6018. [CrossRef] [PubMed]

47. Pochner, K.; Beil, S.; Horn, H.; Blömer, M. Treatment of polymers for subsequent metallization using intense UV radiation or plasma at atmospheric pressure. *Surf. Coat. Technol.* **1997**, *97*, 372–377. [CrossRef]

48. Pascu, M.; Debarnot, D.; Durand, S.; Poncin-Epaillard, F. Surface modification of pvdf by microwave plasma treatment for electroless metallization. In *Plasma Processes and Polymers*; Wiley-VCH Verlag GmbH & Co. KGaA: Weinheim, Germany, 2005; pp. 157–176.

49. Macková, A.; Švorčík, V.; Strýhal, Z.; Pavlík, J. RBS and AFM study of Ag and Au diffusion into pet foils influenced by plasma treatment. *Surf. Interface Anal.* **2006**, *38*, 335–338. [CrossRef]

50. Lin, Y.S.; Liu, H.M. Enhanced adhesion of plasma-sputtered copper films on polyimide substrates by oxygen glow discharge for microelectronics. *Thin Solid Films* **2008**, *516*, 1773–1780. [CrossRef]

51. Li, W.T.; Charters, R.B.; Luther-Davies, B.; Mar, L. Significant improvement of adhesion between gold thin films and a polymer. *Appl. Surf. Sci.* **2004**, *233*, 227–233. [CrossRef]

52. Zille, A.; Fernandes, M.M.; Francesko, A.; Tzanov, T.; Fernandes, M.; Oliveira, F.R.; Almeida, L.; Amorim, T.; Carneiro, N.; Esteves, M.F.; *et al.* Size and aging effects on antimicrobial efficiency of silver nanoparticles coated on polyamide fabrics activated by atmospheric DBD plasma. *ACS Appl. Mater. Interfaces* **2015**, *7*, 13731–13744. [CrossRef] [PubMed]

53. Boyd, R.D.; Kenwright, A.M.; Badyal, J.P.S.; Briggs, D. Atmospheric nonequilibrium plasma treatment of biaxially oriented polypropylene. *Macromolecules* **1997**, *30*, 5429–5436. [CrossRef]
54. Bazaka, K.; Jacob, M.V.; Crawford, R.J.; Ivanova, E.P. Plasma-assisted surface modification of organic biopolymers to prevent bacterial attachment. *Acta Biomater.* **2011**, *7*, 2015–2028. [CrossRef] [PubMed]
55. Poncin-Epaillard, F.; Brosse, J.C.; Falher, T. Cold plasma treatment: Surface or bulk modification of polymer films? *Macromolecules* **1997**, *30*, 4415–4420. [CrossRef]
56. Yasuda, H.; Gazicki, M. Biomedical applications of plasma polymerization and plasma treatment of polymer surfaces. *Biomaterials* **1982**, *3*, 68–77. [CrossRef]
57. Inglis, W.; Sanders, G.H.W.; Williams, P.M.; Davies, M.C.; Roberts, C.J.; Tendler, S.J.B. A simple method for biocompatible polymer based spatially controlled adsorption of blood plasma proteins to a surface. *Langmuir* **2001**, *17*, 7402–7405. [CrossRef]
58. Wang, H.; Kwok, D.T.K.; Wang, W.; Wu, Z.; Tong, L.; Zhang, Y.; Chu, P.K. Osteoblast behavior on polytetrafluoroethylene modified by long pulse, high frequency oxygen plasma immersion ion implantation. *Biomaterials* **2010**, *31*, 413–419. [CrossRef] [PubMed]
59. Oehr, C. Plasma surface modification of polymers for biomedical use. *Nucl. Instrum. Methods Phys. Res. Sect. B* **2003**, *208*, 40–47. [CrossRef]
60. Halfmann, H.; Bibinov, N.; Wunderlich, J.; Awakowicz, P. A double inductively coupled plasma for sterilization of medical devices. *J. Phys. D* **2007**, *40*. [CrossRef]
61. Vesel, A.; Kolar, M.; Stana-Kleinschek, K.; Mozetic, M. Etching rates of blood proteins, blood plasma and polymer in oxygen afterglow of microwave plasma. *Surf. Interface Anal.* **2014**, *46*, 1115–1118. [CrossRef]
62. Stoffels, E.; Flikweert, A.J.; Stoffels, W.W.; Kroesen, G.M.W. Plasma needle: A non-destructive atmospheric plasma source for fine surface treatment of (bio)materials. *Plasma Sources Sci. Technol.* **2002**, *11*. [CrossRef]
63. Herrmann, H.W.; Henins, I.; Park, J.; Selwyn, G.S. Decontamination of chemical and biological warfare (CBW) agents using an atmospheric pressure plasma jet (APPJ). *Phys. Plasmas* **1999**, *6*, 2284–2289. [CrossRef]
64. Lazović, S.; Puač, N.; Miletić, M.; Pavlica, D.; Jovanović, M.; Bugarski, D.; Mojsilović, S.; Maletić, D.; Malović, G.; Milenković, P.; *et al.* The effect of a plasma needle on bacteria in planktonic samples and on peripheral blood mesenchymal stem cells. *New J. Phys.* **2010**, *12*. [CrossRef]
65. Feng, L.; Li, S.; Li, Y.; Li, H.; Zhang, L.; Zhai, J.; Song, Y.; Liu, B.; Jiang, L.; Zhu, D. Super-hydrophobic surfaces: From natural to artificial. *Adv. Mater.* **2002**, *14*, 1857–1860. [CrossRef]
66. Zhang, X.; Shi, F.; Niu, J.; Jiang, Y.; Wang, Z. Superhydrophobic surfaces: From structural control to functional application. *J. Mater. Chem.* **2008**, *18*, 621–633. [CrossRef]
67. Erbil, H.Y.; Demirel, A.L.; Avcı, Y.; Mert, O. Transformation of a simple plastic into a superhydrophobic surface. *Science* **2003**, *299*, 1377–1380. [CrossRef] [PubMed]
68. Puliyalil, H.; Filipič, G.; Cvelbar, U. Recent advances in the methods for designing superhydrophobic surfaces. In *Surface Energy*; Aliofkhazraei, M., Ed.; InTech: Rijeka, Croatia, 2015; pp. 311–335.
69. Toma, M.; Loget, G.; Corn, R.M. Flexible teflon nanocone array surfaces with tunable superhydrophobicity for self-cleaning and aqueous droplet patterning. *ACS Appl. Mater. Interfaces* **2014**, *6*, 11110–11117. [CrossRef] [PubMed]
70. Tsougeni, K.; Vourdas, N.; Tserepi, A.; Gogolides, E.; Cardinaud, C. Mechanisms of oxygen plasma nanotexturing of organic polymer surfaces: From stable super hydrophilic to super hydrophobic surfaces. *Langmuir* **2009**, *25*, 11748–11759. [CrossRef] [PubMed]
71. Korshak, V.V.; Svetlana, V.V. Dependence of thermal stability of polymers on their chemical structure. *Russ. Chem. Rev.* **1968**, *37*, 885. [CrossRef]
72. Taylor, G.N.; Wolf, T.M. Oxygen plasma removal of thin polymer films. *Polym. Eng. Sci.* **1980**, *20*, 1087–1092. [CrossRef]
73. Moss, S.J.; Jolly, A.M.; Tighe, B.J. Plasma oxidation of polymers. *Plasma Chem. Plasma Process.* **1986**, *6*, 401–416. [CrossRef]
74. Zadok, E.; Sialom, B.; Mazur, Y. Oxygen atoms produced by microwave discharge: Reaction with arenes. *Angew. Chem. Int. Ed. Engl.* **1980**, *19*, 1004–1005. [CrossRef]
75. Zadok, E.; Rubinraut, S.; Frolow, F.; Mazur, Y. Reactions of di-, tri-, and hexamethylbenzenes with oxygen(3P) atoms in liquid and on adsorbed phases. *J. Am. Chem. Soc.* **1985**, *107*, 2489–2494. [CrossRef]
76. Xie, L.; Jiao, L.; Dai, H. Selective etching of graphene edges by hydrogen plasma. *J. Am. Chem. Soc.* **2010**, *132*, 14751–14753. [CrossRef] [PubMed]

77. Xiang, H.; Kan, E.; Wei, S.-H.; Whangbo, M.-H.; Yang, J. "Narrow" graphene nanoribbons made easier by partial hydrogenation. *Nano Lett.* **2009**, *9*, 4025–4030. [CrossRef] [PubMed]

78. Kerr, J.A. *Handbook of Bimolecular and Termolecular Gas Reactions*; Taylor & Francis: Boca Raton, FL, USA, 1987.

79. Mogab, C.J.; Adams, A.C.; Flamm, D.L. Plasma etching of Si and SiO$_2$—The effect of oxygen additions to CF$_4$ plasmas. *J. Appl. Phys.* **1978**, *49*, 3796–3803. [CrossRef]

80. Chun, I.; Efremov, A.; Yeom, G.Y.; Kwon, K.-H. A comparative study of CF$_4$/O$_2$/Ar and C$_4$F$_8$/O$_2$/Ar plasmas for dry etching applications. *Thin Solid Films* **2015**, *579*, 136–143. [CrossRef]

81. Hegemann, D.; Brunner, H.; Oehr, C. Plasma treatment of polymers for surface and adhesion improvement. *Nucl. Instrum. Methods Phys. Res. Sect. B* **2003**, *208*, 281–286. [CrossRef]

82. Wohlfart, E.; Fernández-Blázquez, J.P.; Knoche, E.; Bello, A.; Pérez, E.; Arzt, E.; del Campo, A. Nanofibrillar patterns by plasma etching: The influence of polymer crystallinity and orientation in surface morphology. *Macromolecules* **2010**, *43*, 9908–9917. [CrossRef]

83. Junkar, I.; Cvelbar, U.; Vesel, A.; Hauptman, N.; Mozetič, M. The role of crystallinity on polymer interaction with oxygen plasma. *Plasma Process. Polym.* **2009**, *6*, 667–675. [CrossRef]

84. Chernomordik, B.D.; Russel, H.B.; Cvelbar, U.; Jasinski, J.B.; Kumar, V.; Deutsch, T.; Sunkara, M.K. Photoelectrochemical activity of as-grown, α-Fe$_2$O$_3$ nanowire array electrodes for water splitting. *Nanotechnology* **2012**, *23*. [CrossRef] [PubMed]

85. Lazović, S.; Puač, N.; Spasić, K.; Malović, G.; Cvelbar, U.; Mozetič, M.; Radetić, M.; Petrović, Z.L. Plasma properties in a large-volume, cylindrical and asymmetric radio-frequency capacitively coupled industrial-prototype reactor. *J. Phys. D Appl. Phys.* **2013**, *46*. [CrossRef]

86. Olde Riekerink, M.B.; Terlingen, J.G.A.; Engbers, G.H.M.; Feijen, J. Selective etching of semicrystalline polymers: CF$_4$ gas plasma treatment of poly(ethylene). *Langmuir* **1999**, *15*, 4847–4856. [CrossRef]

87. Nair, L.G.; Mahapatra, A.S.; Gomathi, N.; Joseph, K.; Neogi, S.; Nair, C.P.R. Radio frequency plasma mediated dry functionalization of multiwall carbon nanotube. *Appl. Surf. Sci.* **2015**, *340*, 64–71. [CrossRef]

88. Black, C.T.; Guarini, K.W.; Milkove, K.R.; Baker, S.M.; Russell, T.P.; Tuominen, M.T. Integration of self-assembled diblock copolymers for semiconductor capacitor fabrication. *Appl. Phys. Lett.* **2001**, *79*, 409–411. [CrossRef]

89. Chuang, V.P.; Ross, C.A.; Gwyther, J.; Manners, I. Self-assembled nanoscale ring arrays from a polystyrene-*b*-polyferrocenylsilane-*b*-poly(2-vinylpyridine)triblock terpolymer thin film. *Adv. Mater.* **2009**, *21*, 3789–3793. [CrossRef]

90. Lammertink, R.G.H.; Hempenius, M.A.; van den Enk, J.E.; Chan, V.Z.H.; Thomas, E.L.; Vancso, G.J. Nanostructured thin films of organic–organometallic block copolymers: One-step lithography with poly(ferrocenylsilanes) by reactive ion etching. *Adv. Mater.* **2000**, *12*, 98–103.

91. Jung, Y.S.; Ross, C.A. Orientation-controlled self-assembled nanolithography using a polystyrene—Polydimethylsiloxane block copolymer. *Nano Lett.* **2007**, *7*, 2046–2050. [CrossRef] [PubMed]

92. Kim, S.Y.; Nunns, A.; Gwyther, J.; Davis, R.L.; Manners, I.; Chaikin, P.M.; Register, R.A. Large-area nanosquare arrays from shear-aligned block copolymer thin films. *Nano Lett.* **2014**, *14*, 5698–5705. [CrossRef] [PubMed]

93. Mozetič, M. Controlled oxidation of organic compounds in oxygen plasma. *Vacuum* **2003**, *71*, 237–240. [CrossRef]

94. Panda, S.; Economou, D.J.; Meyyappan, M. Effect of metastable oxygen molecules in high density power-modulated oxygen discharges. *J. Appl. Phys.* **2000**, *87*, 8323–8333. [CrossRef]

95. Dai, L.; Griesser, H.J.; Mau, A.W.H. Surface modification by plasma etching and plasma patterning. *J. Phys. Chem. B* **1997**, *101*, 9548–9554. [CrossRef]

96. Chen, F.F.; Smith, M.D. Plasma. In *Van Nostrand's Scientific Encyclopedia*; John Wiley & Sons, Inc.: Hoboken, NJ, USA, 2005. [CrossRef]

97. Donnelly, V.M.; Kornblit, A. Plasma etching: Yesterday, today, and tomorrow. *J. Vac. Sci. Technol. A* **2013**, *31*. [CrossRef]

98. Mozetič, M.; Ostrikov, K.; Ruzic, D.N.; Curreli, D.; Cvelbar, U.; Vesel, A.; Primc, G.; Leisch, M.; Jousten, K.; Malyshev, O.B.; *et al.* Recent advances in vacuum sciences and applications. *J. Phys. D* **2014**, *47*. [CrossRef]

99. Kontziampasis, D.; Constantoudis, V.; Gogolides, E. Plasma directed organization of nanodots on polymers: Effects of polymer type and etching time on morphology and order. *Plasma Process. Polym.* **2012**, *9*, 866–872. [CrossRef]

100. Vourdas, N.; Kontziampasis, D.; Kokkoris, G.; Constantoudis, V.; Goodyear, A.; Tserepi, A.; Cooke, M.; Gogolides, E. Plasma directed assembly and organization: Bottom-up nanopatterning using top-down technology. *Nanotechnology* **2010**, *21*. [CrossRef] [PubMed]

101. Puliyalil, H.; Cvelbar, U.; Filipic, G.; Petric, A.D.; Zaplotnik, R.; Recek, N.; Mozetic, M.; Thomas, S. Plasma as a tool for enhancing insulation properties of polymer composites. *RSC Adv.* **2015**, *5*, 37853–37858. [CrossRef]

102. El-Aguizy, T.A.; Jeong, J.-H.; Jeon, Y.-B.; Li, W.Z.; Ren, Z.F.; Kim, S.-G. Transplanting carbon nanotubes. *Appl. Phys. Lett.* **2004**, *85*, 5995–5997. [CrossRef]

103. Che, Y.; Wang, C.; Liu, J.; Liu, B.; Lin, X.; Parker, J.; Beasley, C.; Wong, H.S.P.; Zhou, C. Selective synthesis and device applications of semiconducting single-walled carbon nanotubes using isopropyl alcohol as feedstock. *ACS Nano* **2012**, *6*, 7454–7462. [CrossRef] [PubMed]

104. Hou, P.-X.; Li, W.-S.; Zhao, S.-Y.; Li, G.-X.; Shi, C.; Liu, C.; Cheng, H.-M. Preparation of metallic single-wall carbon nanotubes by selective etching. *ACS Nano* **2014**, *8*, 7156–7162. [CrossRef] [PubMed]

105. Zhang, G.; Qi, P.; Wang, X.; Lu, Y.; Li, X.; Tu, R.; Bangsaruntip, S.; Mann, D.; Zhang, L.; Dai, H. Selective etching of metallic carbon nanotubes by gas-phase reaction. *Science* **2006**, *314*, 974–977. [CrossRef] [PubMed]

106. Di, C.-A.; Wei, D.; Yu, G.; Liu, Y.; Guo, Y.; Zhu, D. Patterned graphene as source/drain electrodes for bottom-contact organic field-effect transistors. *Adv. Mater.* **2008**, *20*, 3289–3293. [CrossRef]

107. Liao, Z.; Wan, Q.; Liu, H.; Tang, Q. Realization of size controllable graphene micro/nanogap with a micro/nanowire mask method for organic field-effect transistors. *Appl. Phys. Lett.* **2011**, *99*. [CrossRef]

108. Jiang, L.; Fan, Z. Design of advanced porous graphene materials: From graphene nanomesh to 3d architectures. *Nanoscale* **2014**, *6*, 1922–1945. [CrossRef] [PubMed]

109. Song, X.; Hu, J.; Zeng, H. Two-dimensional semiconductors: Recent progress and future perspectives. *J. Mater. Chem. C* **2013**, *1*, 2952–2969. [CrossRef]

110. Zang, J.; Ryu, S.; Pugno, N.; Wang, Q.; Tu, Q.; Buehler, M.J.; Zhao, X. Multifunctionality and control of the crumpling and unfolding of large-area graphene. *Nat. Mater.* **2013**, *12*, 321–325. [CrossRef] [PubMed]

111. Zeng, Z.; Huang, X.; Yin, Z.; Li, H.; Chen, Y.; Li, H.; Zhang, Q.; Ma, J.; Boey, F.; Zhang, H. Fabrication of graphene nanomesh by using an anodic aluminum oxide membrane as a template. *Adv. Mater.* **2012**, *24*, 4138–4142. [CrossRef] [PubMed]

112. Kim, M.; Safron, N.S.; Han, E.; Arnold, M.S.; Gopalan, P. Fabrication and characterization of large-area, semiconducting nanoperforated graphene materials. *Nano Lett.* **2010**, *10*, 1125–1131. [CrossRef] [PubMed]

113. Ding, J.; Du, K.; Wathuthanthri, I.; Choi, C.-H.; Fisher, F.T.; Yang, E.-H. Transfer patterning of large-area graphene nanomesh via holographic lithography and plasma etching. *J. Vac. Sci. Technol. Technol. B* **2014**, *32*. [CrossRef]

114. Sinitskii, A.; Tour, J.M. Patterning graphene through the self-assembled templates: Toward periodic two-dimensional graphene nanostructures with semiconductor properties. *J. Am. Chem. Soc.* **2010**, *132*, 14730–14732. [CrossRef] [PubMed]

115. Liang, X.; Jung, Y.-S.; Wu, S.; Ismach, A.; Olynick, D.L.; Cabrini, S.; Bokor, J. Formation of bandgap and subbands in graphene nanomeshes with sub-10 nm ribbon width fabricated via nanoimprint lithography. *Nano Lett.* **2010**, *10*, 2454–2460. [CrossRef] [PubMed]

116. Paul, R.K.; Badhulika, S.; Saucedo, N.M.; Mulchandani, A. Graphene nanomesh as highly sensitive chemiresistor gas sensor. *Anal. Chem.* **2012**, *84*, 8171–8178. [CrossRef] [PubMed]

117. Fatyeyeva, K.; Dahi, A.; Chappey, C.; Langevin, D.; Valleton, J.-M.; Poncin-Epaillard, F.; Marais, S. Effect of cold plasma treatment on surface properties and gas permeability of polyimide films. *RSC Adv.* **2014**, *4*, 31036–31046. [CrossRef]

118. Zarshenas, K.; Raisi, A.; Aroujalian, A. Surface modification of polyamide composite membranes by corona air plasma for gas separation applications. *RSC Adv.* **2015**, *5*, 19760–19772. [CrossRef]

119. Chen, S.-H.; Chuang, W.-H.; Wang, A.A.; Ruaan, R.-C.; Lai, J.-Y. Oxygen/nitrogen separation by plasma chlorinated polybutadiene/polycarbonate composite membrane. *J. Membr. Sci.* **1997**, *124*, 273–281. [CrossRef]

120. Chen, Z.; Tan, X. Monolithic fabrication of ionic polymer–metal composite actuators capable of complex deformation. *Sens. Actuators A* **2010**, *157*, 246–257. [CrossRef]

121. Nikhil Dilip, P.; Mark, S.M. An investigation of the fracturing process in nitrogen-doped multiwalled carbon nanotubes (N-MWCNTs). Evidence for directional unzipping. *Mater. Res. Express* **2014**, *1*. [CrossRef]

122. Jiao, L.; Zhang, L.; Wang, X.; Diankov, G.; Dai, H. Narrow graphene nanoribbons from carbon nanotubes. *Nature* **2009**, *458*, 877–880. [CrossRef] [PubMed]

123. Poulsen, R.G. Plasma etching in integrated circuit manufacture—A review. *J. Vac. Sci. Technol.* **1977**, *14*, 266–274. [CrossRef]

124. Vitale, S.A.; Berry, S. Etching selectivity of indium tin oxide to photoresist in high density chlorine- and ethylene-containing plasmas. *J. Vac. Sci. Technol. B* **2013**, *31*. [CrossRef]

125. Laermer, F.; Urban, A. Challenges, developments and applications of silicon deep reactive ion etching. *Microelectron. Eng.* **2003**, *67–68*, 349–355. [CrossRef]

126. Liao, Q.; Xu, Z.; Zhong, X.; Dang, W.; Shi, Q.; Zhang, C.; Weng, Y.; Li, Z.; Fu, H. An organic nanowire waveguide exciton-polariton sub-microlaser and its photonic application. *J. Mater. Chem. C* **2014**, *2*, 2773–2778. [CrossRef]

127. Garcia-Frutos, E.M. Small organic single-crystalline one-dimensional micro- and nanostructures for miniaturized devices. *J. Mater. Chem. C* **2013**, *1*, 3633–3645. [CrossRef]

128. Fang, H.; Wu, W.; Song, J.; Wang, Z.L. Controlled growth of aligned polymer nanowires. *J. Phys. Chem. C* **2009**, *113*, 16571–16574. [CrossRef]

129. Chen, Y.; Xu, Z.; Gartia, M.R.; Whitlock, D.; Lian, Y.; Liu, G.L. Ultrahigh throughput silicon nanomanufacturing by simultaneous reactive ion synthesis and etching. *ACS Nano* **2011**, *5*, 8002–8012. [CrossRef] [PubMed]

130. Chandran, N.; Chandran, S.; Maria, H.J.; Thomas, S. Compatibilizing action and localization of clay in a polypropylene/natural rubber (PP/NR) blend. *RSC Adv.* **2015**, *5*, 86265–86273. [CrossRef]

131. Jyotishkumar, P.; Koetz, J.; Tiersch, B.; Strehmel, V.; Özdilek, C.; Moldenaers, P.; Hässler, R.; Thomas, S. Complex phase separation in poly(acrylonitrile–butadiene–styrene)-modified epoxy/4,4′-diaminodiphenyl sulfone blends: Generation of new micro- and nanosubstructures. *J. Phys. Chem. B* **2009**, *113*, 5418–5430. [CrossRef] [PubMed]

132. Kunaver, M.; Mozetič, M.; Klanjšek-Gunde, M. Selective plasma etching of powder coatings. *Thin Solid Films* **2004**, *459*, 115–117. [CrossRef]

133. Kunaver, M.; Klanjsek-Gunde, M.; Mozetic, M.; Kunaver, M.; Hrovat, A. The degree of dispersion of pigments in powder coatings and the origin of some surface defects. *Surf. Coat. Int. Part B* **2003**, *86*, 175–179. [CrossRef]

134. Mozetič, M.; Zalar, A.; Panjan, P.; Bele, M.; Pejovnik, S.; Grmek, R. A method of studying carbon particle distribution in paint films. *Thin Solid Films* **2000**, *376*, 5–8. [CrossRef]

135. Salalha, W.; Dror, Y.; Khalfin, R.L.; Cohen, Y.; Yarin, A.L.; Zussman, E. Single-walled carbon nanotubes embedded in oriented polymeric nanofibers by electrospinning. *Langmuir* **2004**, *20*, 9852–9855. [CrossRef] [PubMed]

136. Ponnamma, D.; Sadasivuni, K.K.; Grohens, Y.; Guo, Q.; Thomas, S. Carbon nanotube based elastomer composites—An approach towards multifunctional materials. *J. Mater. Chem. C* **2014**, *2*, 8446–8485. [CrossRef]

137. Cvelbar, U.; Pejovnik, S.; Mozetiè, M.; Zalar, A. Increased surface roughness by oxygen plasma treatment of graphite/polymer composite. *Appl. Surf. Sci.* **2003**, *210*, 255–261. [CrossRef]

138. Malara, F.; Manca, M.; De Marco, L.; Pareo, P.; Gigli, G. Flexible carbon nanotube-based composite plates as efficient monolithic counter electrodes for dye solar cells. *ACS Appl. Mater. Interfaces* **2011**, *3*, 3625–3632. [CrossRef] [PubMed]

139. Zhang, Y.; Li, J.; An, G.; He, X. Highly porous SnO_2 fibers by electrospinning and oxygen plasma etching and its ethanol-sensing properties. *Sens. Actuators B* **2010**, *144*, 43–48. [CrossRef]

140. Ramgir, N.; Datta, N.; Kaur, M.; Kailasaganapathi, S.; Debnath, A.K.; Aswal, D.K.; Gupta, S.K. Metal oxide nanowires for chemiresistive gas sensors: Issues, challenges and prospects. *Colloids Surf. A* **2013**, *439*, 101–116. [CrossRef]

141. Bandgar, D.K.; Navale, S.T.; Nalage, S.R.; Mane, R.S.; Stadler, F.J.; Aswal, D.K.; Gupta, S.K.; Patil, V.B. Simple and low-temperature polyaniline-based flexible ammonia sensor: A step towards laboratory synthesis to economical device design. *J. Mater. Chem. C* **2015**, *3*, 9461–9468. [CrossRef]

142. Han, J.-W.; Kim, B.; Li, J.; Meyyappan, M. Carbon nanotube based humidity sensor on cellulose paper. *J. Phys. Chem. C* **2012**, *116*, 22094–22097. [CrossRef]

143. Raghu, M.; Suresh, R.; Vijay, P.S. MWCNT–polymer composites as highly sensitive and selective room temperature gas sensors. *Nanotechnology* **2011**, *22*. [CrossRef]

144. Ernest, M.; Jahir, O.; Cecilia, J.-J.; Ana, B.G.-G.; Ana, C.; Laura, M.L.; César, F.-S. Scalable fabrication of immunosensors based on carbon nanotube polymer composites. *Nanotechnology* **2008**, *19*. [CrossRef]
145. Sullalti, S.; Colonna, M.; Berti, C.; Fiorini, M.; Karanam, S. Effect of phosphorus based flame retardants on UL94 and comparative tracking index properties of poly(butylene terephthalate). *Polym. Degrad. Stab.* **2012**, *97*, 566–572. [CrossRef]

MDPI AG
St. Alban-Anlage 66
4052 Basel, Switzerland
Tel. +41 61 683 77 34
Fax +41 61 302 89 18
http://www.mdpi.com

Nanomaterials Editorial Office
E-mail: nanomaterials@mdpi.com
http://www.mdpi.com/journal/nanomaterials

www.ingramcontent.com/pod-product-compliance
Lightning Source LLC
Chambersburg PA
CBHW051857210326
41597CB00033B/5936